'Bella is one of the most inspiring people I've ever met. She has an eloquence and lucidity that is timeless, partnered with a powerful combination of positivity and belief. Bella and her fellow young voices are the best chance our planet has. I've seen Bella bring a theatre full of academics and conservationists to tears, and then to their feet. This young woman has an oratory gift that any storyteller would kill for, and a passion and energy that is infectious and dazzling. Bella believes she can change the world, and I believe her. *The Children of the Anthropocene* is a remarkable and important book'
Steve Backshall, naturalist, broadcaster and author

'From the Amazon rainforests to the beaches of Mumbai, the city streets of the US and the farms of Europe, Bella Lack hears from young people at the sharp end of the environmental crisis who are challenging the economic and political system that has led us to where we are now, with a deeply damaged world and facing a climate and ecological catastrophe. This book is so much more than a record of what's gone wrong, it's an inspirational manifesto for change. As a passionate campaigner herself, Bella is the perfect guide'
Caroline Lucas, former leader of The Green Party

'A visionary statement for the future, from a brilliant young person who hopes the planet will be there to enjoy it. Pragmatic, positive and beautifully written'
Ben Macdonald, award-winning conservation writer, wildlife TV producer and naturalist

'Astute, erudite and crystalline, Bella writes with visionary clarity and passion. It was a pleasure to be interviewed by Bella for *The Children of the Anthropocene*, she questions everything with intelligence, grace and humour. It's a wonderful book'
Dara McAnulty, award-winning author of
Diary of a Young Naturalist

'In *The Children of the Anthropocene* you will hear inspiring stories from all over the world. Stories from the most affected people in the most affected areas. Unheard but not voiceless, their stories must be told. So, let's pass the mic'
Greta Thunberg

'An urgent, thought-provoking and beautifully written book from a brilliant young conservationist. Bella vividly brings the wonders of nature to life on the page, showcasing the importance of diverse human stories in the collective fight to protect our planet in the face of the environmental crisis. We must stop and listen to these inspiring young people from around the globe. Extraordinarily moving, wild and engaging – the book of the moment'
Mary Robinson, former president of
Ireland and author of *Climate Justice*

'A passionate call for change. Hope-filled and fascinating, this book has inspired me to do more'
Dave Goulson, author of the *Sunday Times* bestselling
The Garden Jungle and *A Sting in the Tale*

'A remarkably lucid and insightful account of the problem with our relationship with the natural world, and how we can save it and ourselves'
Liz Bonnin, natural history and
environmental broadcaster

'Profound wisdom from a brilliant young mind – Bella's view of our troubled planet is enthralling and shocking, inspiring and enchanting. She articulates the exhilarating and fresh perspectives of a rising generation determined to turn things round. Clear-headed about the evidence and passionate about the answers, this book offers something remarkable: real hope'
David Shukman, environmental journalist and writer

'Bella Lack has woven a beautiful offering to the world in her book *The Children of the Anthropocene*. A tapestry of stories and facts, encouragement and holding to account; she brings to life both the possibility of change and the longing of the generation made to face the consequences of our diseased way of living. Set aside any prejudice about age – whether assumptions about naivety or projections of hope on to our young people, Bella is inviting a deep collaboration among all of us, bringing whatever gifts we can offer to this great collective endeavour for transformation'
Dr Gail Bradbrook, Extinction Rebellion co-founder

'Thought-provoking without being preachy this is a really serious and helpful book cleverly using the personal stories of people directly affected by some part of climate change. All from a young person who is positive about finding a better way for us all to live in the future'
David Lindo, The Urban Birder

'*The Children of the Anthropocene* is not only a book of pain and defiance, resilience and love as Bella Lack writes, it is by and for courageous, compassionate and dedicated young people on the frontlines of the climate emergency. Their too-often-ignored stories offer hope, power and inspiration that we will realize a fairer, greener and healthier world for all'
Jennifer Morgan, Executive Director,
Greenpeace International

ABOUT THE AUTHOR

Bella Lack is an eighteen-year-old conservationist and environmental activist. She is an ambassador for the Born Free Foundation, STAE, RSPCA and the Jane Goodall Institute. Bella spoke at Chris Packham's The People's Walk for Wildlife, the Illegal Wildlife Trade Conference in 2018, and she delivered a TEDx talk in Brighton in 2019.

Bella creates short films and uses social media to educate and inspire as many people as possible to protect the natural world. Since 2019, she has been working on a feature-length documentary with primatologist Jane Goodall, called *Animal.*

She has been interviewed on Sky, ITV, Channel 4, CGTN in China and has also made a short documentary for BBC Three. She has shared the stage with the likes of Steve Backshall and Chris Packham and helped to create *The People's Manifesto for Wildlife*. In the year running up to COP26, she worked alongside Alok Sharma, co-chairing the civil society and youth advisory council.

The Children of the Anthropocene

Stories from the Young People at the Heart of the Climate Crisis

BELLA LACK

PENGUIN LIFE

AN IMPRINT OF

PENGUIN BOOKS

PENGUIN LIFE

UK | USA | Canada | Ireland | Australia
India | New Zealand | South Africa

Penguin Life is part of the Penguin Random House group of companies whose addresses can be found at global.penguinrandomhouse.com

First published 2022
001

Set in 12.5/14.75pt Garamond MT Std
Typeset by Jouve (UK), Milton Keynes
Printed and bound in Great Britain by Clays Ltd, Elcograf S.p.A.

The authorized representative in the EEA is Penguin Random House Ireland, Morrison Chambers, 32 Nassau Street, Dublin D02 YH68

A CIP catalogue record for this book is available from the British Library

ISBN: 978–0–241–50108–5

www.greenpenguin.co.uk

To those young people who don't see a decimated planet and allow that to silence them, but rather see a transformed future and allow that to empower them. Thank you.

Contents

The Children of the Anthropocene

A Foreword by Greta Thunberg

We are right now in the beginning of a global climate and ecological emergency that affects all of us.

But not everyone is suffering its consequences equally. Where I come from, we often speak of the climate crisis like something that will affect our children and grandchildren in a distant future. But what we seem to forget over and over is that people are dying due to its consequences already today.

For decades, people – especially youth – have been tirelessly fighting for climate and ecological justice. While we wait for the people in power to act, in every city, every community and every country, young people are speaking up and taking action. And I'm so proud to be fighting alongside them.

Our activism takes many different forms, but we are working together towards the same goal: changing political and economic systems to prioritize people and the planet.

In *The Children of the Anthropocene*, you will hear inspiring stories from all over the world. Stories from the most affected people in the most affected areas. Unheard but not voiceless, their stories must be told.

So let's pass the mic.

Introduction

We are addicted to stories. They are the cornerstone of our civilization. Stories characterize humans like flight characterizes birds, like petals characterize flowers. They're the essence of our species and are the most powerful form of human communication.

I realized this in March 2020.

It didn't look like the end of the world, although many alarmists on Twitter assured me it was. I watched the world carefully from the window. The morning was damp and steaming like a newborn calf. A few milky aeroplane contrails braided across the blue sky. The ambient hum of London was muffled and remote for the first time in my life. Later, when the spring mist began to clear, I pulled on a large sweater and walked in the park. It was that kind of magical day when spring turns from an inkling into an explosive emergence of blooms, buds and plump bees. A day when daffodils thrust their heads into the sunlight and croaking frogs serenade each other in the green ponds. The birds were singing a little louder, the sun shining a little brighter, and the grey British winter was thawing out; life was renewed. The air seemed to be thrumming with activity, and yet for once, it wasn't humans. When I did see a fellow *Homo sapiens*, we politely began to veer away from each other a few metres in advance, to keep the angle subtle but large enough to ensure we maintained the obligatory two metres (to be fair, this type of behaviour is not unusual for British people on a normal day).

Across the world, we humans sat in our homes, each of us neatly packaged within our four walls. In a few short weeks, our species had retreated from the constant, gnawing labour of routine and become untethered. The machine of humanity had ground to a halt and our horizons had shrunk from the global neighbourhood to the achingly familiar sight of the inside of our front doors.

What surprised me most was the ease and rapidity with which major capital cities had transformed from bustling metropolises to hauntingly hollow ghost towns. The how and why was obvious, though. It wasn't just government warnings and regulations keeping us contained, but the stories that cascaded down our social media feeds as well. We heard about a sick mother in Wuhan, a terrified couple marooned on an infected cruise ship. We read about our friend's grandad in hospital, our prime minister in the ICU, and an Italian nurse exhausted and bruised from hours on shift. These stories personalized and reinforced the severity of the virus for the global population more than any statistic could ever do.

In 2020, stories, not just statistics, mobilized the world to care about the coronavirus. Information about the state of our planet is often complex, unengaging and absent from wider public discourse. Linguistic analysis even found that the most recent IPCC (Intergovernmental Panel on Climate Change) report was less readable than seminal papers by Albert Einstein, making communication of the reality of the climate crisis even harder. So that's why this is not a book of cold, hard statistics and science. Instead, it is a book of pain and defiance, resilience and love.

Science has repeatedly predicted outcomes such as:

- 7 million people will die due to air pollution every year
- sea level could rise 34 cm by 2050
- 1 million species will be threatened with extinction
- by 2050, the oceans could contain more plastic than fish

Now it's time for these statistics to be finally given a face.

When we hear scientific terms such as 'ocean acidification', we should all feel that these are some of the biggest words in the world, because this is one of the biggest chemical changes to afflict earth in the last fifty million years. Strangely, though, we don't feel much at all when we hear these words, because they have no personal weight to them and are disconnected from our everyday lives. They are not entangled in our memory or embedded in our emotion. We don't feel they are connected to the music we like, the people we love, the Second World War, the Beatles or the first time we rode a bike. They are just words, a few straggles of ink etched onto a page, and yet the lives of many people have been transformed and torn apart by these words and the environmental crisis at large.

Our whimsical Western fantasies of perpetual growth, development and endless energy work for a small minority, but plunge much of the population into solastalgia. Solastalgia is best described as 'the lived experience of negatively perceived environmental change'. In other words, it's when a home becomes unhomely in front of your eyes. It's when 'your endemic sense of place has been violated'. It is when you are forced to lament the landscape you once loved. It's nostalgia for a place that once was, and a fear for the future to come.

Living in England means I lament the loss of nightingales, red squirrels and butterflies, but my life remains fundamentally the same whether I see a high brown fritillary in summer or not. In the UK, many of us welcome the warmer summers and milder winters. However, what has changed most immediately is our psychological awareness that disaster looms on the intangible horizon. For a vast number of young people across the world, the games, naïvety, whims and indulgences that should characterize their childhood are having to be cast aside. Where many of us in the West are saying, 'we must protect elephants so our grandchildren won't have to live in a world without them', young people at the heart of the crisis around the world are saying, 'we must protect our planet so that I don't wake up unable to breathe because of air pollution, or riddled with cancer because of the plastic and chemicals in my water source, or displaced as the forest where I live has been cut down'.

Across the world, a plethora of people are being used as currency to buy certain people more cars, bigger televisions and faster flights. So, for a moment, let's take a step back and listen to the stories of those young people at the heart of the crisis who are directly impacted by these issues. Let's slow down and *really* listen, because none of these stories are yet complete and their endings can still be changed. It is the role of humanity, of you and me, to confront society's sluggish resistance to change and to imagine a different story for the future. For those who will follow. Imagination is like memories that haven't hardened yet, and we must use this to sculpt and shape our future before it cements into the annals of history.

This book bears witness to the unfolding of history. It holds up a looking glass to the events occurring in the world and will hopefully magnify your desire to do something not only for the people who have shared their stories, but also for the places they belong to, and for the other species that roam these lands.

Prologue

Us and Them

'They'll kill you without a second thought,' she warned me. I was sitting in the violet light of dusk in the Ecuadorian jungle, on the phone to my mum after over a week of being without signal. I was excited and nervous, watching that delicious heavy gloom you only find somewhere remote settle over the treetops, knowing that something big was going to happen the next morning.

We were camping in the Chocó cloud rainforest on a conservation and scientific expedition to find and describe new species – four young people, including me, and three adults. We had been trekking through the reserve we had spent two years fundraising for and working to protect, and this was the first time I had been there in person.

On one of his solitary escapades, twenty-four hours earlier, Marco, a fellow member of the youth council and an Ecuadorian local who knew his way around the jungle with his eyes closed, had discovered a rugged trail snaking close to our makeshift camp. He followed it, machete gripped in one hand, the other twisting off lianas and pulling flat, fat leaves from his face as they whipped him on the thin path. When he returned, I was at the camp with the five other expedition members. He gathered us round and opened his palm. I didn't see what he was holding at first, but I heard the sharp intakes of breath around me and saw our expedition leader, Javier, anxiously sweep his hair back from his face and recoil in shock. I knew something was wrong.

After some communication, and using my fragments of Spanish, I understood that the shard of yellow rock he held in his hand was a mixture of gold and pyrite. Marco had found an illegal mining site, on a protected reserve. The area we had worked tirelessly to establish as a reserve over the last two years had been encroached upon and was now being targeted by mining companies. The miners had decimated over three kilometres of an ecologically significant gorge – habitat for several threatened and endangered species – and by the looks of it, they were still in the area. Their tracks were fresh.

There was a moment of deadly silence. It felt like our hopes and ambitions for the incredible patch of wilderness we were sitting in had been liquidated. Then everyone's resolve tangibly hardened. Eyes narrowed, voices lifted, plans and ideas were shot back and forth. We sent up a drone, scouring the dense jungle for several kilometres before we found the miners' base camp. That night, we retreated to a safer location further away. The next morning, before sunrise, we confronted them.

It took us two attempts, but eventually they told us who they were working for, and we retreated once again with that piece of crucial information tucked away and stored for the battle that was to come.

The mining companies often establish themselves in an area and will give the local community an ultimatum of being displaced or being employed. What I learnt in those tense twenty-four hours was that the 'us and them' mentality we have in life, and in our approach to the climate crisis – e.g. youth vs adults, conservationists vs miners – doesn't work. The priority for everyone involved in a situation is almost always to survive; life is a desperate scramble to look after ourselves, our friends, families and communities. The one thing we knew about the miners was that they were destroying the reserve, so it was easy to construct a whole malicious persona about them in our minds. However, behind that was likely a group of people simply wanting to look after their families, who had not been given the same opportunities and resources as many others had.

When given the chance, and armed with the knowledge needed, people will often change. The climate crisis is not the fault of society at large, but of a relatively small band of companies and their powerful PR campaigns. When large companies, ideas and ideologies move your way, it's easy to latch on and allow the social current to whisk you towards the land of conformity. Therefore, this book is not directed towards any one individual, demographic, generation or group of people. It is, rather, a call for change that may be heard and heeded by any young person, activist, hunter, lawyer, politician or artist. I really believe that change is possible for anyone if they make the effort to step outside their paradigm and re-evaluate the way they exist on and treat our wonderful planet.

Chapter 1

A Species Who Won't Stop Consuming

Plastic, overconsumption and circularity

25 November 2019

As we stood there on that rubbish-strewn apocalyptic landscape shrouded in human waste of all elements and kinds, I felt almost a bit relieved. It was like an intoxicating devastation. The problem facing us felt so big and so insurmountable that I wondered whether I should stop with all the campaigning, speeches and activism and just enjoy my teenage years whilst they lasted. I was seventeen in two days and I was in Mumbai, on one of the most polluted beaches in the world, not feeling much at all.

Looking back, I realize this numbness was a very real, very human response. In fact, it's a response so common that it could be exactly why we are in such a dire environmental situation yet still lacking the motivation to act. This is called psychic numbing. If I tell you about my dog dying, you will feel empathy and sadness for my loss. But you can't feel empathy and sadness for every dog death in the world, or you would constantly be in a state of mourning for the many dogs dying every day (sorry we had to go there; around 1.2 million puppies were born today!).

This is the same infuriating paradox that keeps people and politicians from intervening in humanitarian crises. The way we value a person's life declines precipitously with numbers. A story about an individual victim speaks to our heart, but a dry statistic about millions speaks to our head. This is why you've heard about Cecil the lion and Harambe the gorilla, but not the millions of animals killed by humans every day in slaughterhouses. People won't act unless they feel they can really change something, and that's the essence of it – efficacy. People would rather do nothing than do something that feels ineffective. Standing there with my feet buried in humanity's waste, it certainly felt ineffective to refuse a plastic straw, or to take a reusable cup to the coffee shop.

Plastic has for a while been the poster child of environmentalism. It is now cowering ever so slightly forgotten in the corner as its bigger cousin, climate change, takes centre stage. But plastic is a good entry point into understanding the 'Anthropocene' of this book's title. The Anthropocene is the period of time in which the impact of humanity has become so profound that a new geological epoch needs to be declared, meaning we have become a geophysical force on a planetary scale. Scientists think hypothetical future geologists will be able to define the era by a layer of 'plastiglomerate'. In other words, that disposable coffee cup you used three years ago could be your most durable legacy.

You, I and all the other thin-skinned, individually fragile *Homo sapiens* manage to produce more than 300 million tonnes of one of the most imperishable materials on the planet every single year. This is the paradox of durability. We want products to be hardy, but not too hardy. We want them to last, but not last for too long.

Plastic is not the problem in itself, unfortunately; we

are (something that will be a recurring theme throughout this book). Our culture allows us to believe that throwing high volumes of stuff away is the inevitable reality of living, but even the term 'throw away' is misleading. There is no 'away', only 'away from me'. Much of the plastic around today will take up to a thousand years to decompose, although 'decompose' is misleading too, because it suggests absence, when in reality it may be gone from our vision but will still lurk around as an ocean cloud or a terrestrial layer of microplastics that will forever memorialize this moment in natural history.

Plastic doesn't decide for itself to stick around for so long, or lodge itself in the stomachs of seabirds, so for fear of blaming it for our own wrongdoing, we first need to discuss consumerism. It's very hard to objectively analyse and distance yourself from something that has been the beating heart of our culture for generations, but we must try if we're going to understand ourselves and our past enough to create a better future.

Imagine the global corporate consumer monoculture like the *Titanic* on a collision course with the natural world (the iceberg). On the top deck there is a deadly and desperate illusion of normality, whilst the bottom compartments are being churned and pounded by icy torrents of water. Wherever you are on the ship, however, the question is not *if* you'll sink, but when. Unless, that is, you take steps to protect yourself and others, investing in interdependence rather than independence. The top deck is much of the Western world and our bunker 'us vs them' mentality. Our sense of security is bolstered by all this stuff we can easily consume, but it's fleeting and fulfils no one. In fact, it's like those seabirds with their plastic. We keep consuming more and more in the hope that it will fill us up, but our hunger becomes more ravenous, more desperate and insatiable. To stay active, the consumer society must always promise satisfaction, yet never ultimately deliver it.

It's very human. That's something else you'll realize as we embark on this messy journey through the best and worst of human capabilities. Often chaos, consumption and destruction are states that humans revert to without realizing. Changing ourselves is going to be hard, but as with revising for an exam or training for a marathon, the reward will be great and long-lasting.

It is enticing, though, this constant overbuying, overeating hyperconsumption. The reason for this is that advertisers and corporations know you very, very well. Cattle run together. Birds flock together. Bees swarm together. Humans do the same; we copy others all the time. It is one of the reasons we've managed to accomplish so much in our time on this earth, and when everybody is doing something, this consensus gives us permission to do it more and more.

The term 'consumerism' does not simply refer to the omnipresence of advertising, or the fast-food chains that litter the high street, but anything connected to the overarching idea that to be better and more successful humans, we have to have more stuff. This is the saddest aspect of the illusion we're all under. Often brands will tap into our desire to belong in order to entice us to buy their products. For example, Doritos tortilla chips created an ad that linked the product to the idea of friendship and companionship. Subconsciously, we begin to believe that one particular brand of triangular crisps is going to fulfil a loneliness within us. In fact, according to a study from the Money and Mental Health Policy Institute, a UK charity, nine out of ten people who struggle with mental health issues spend more when they do not feel well. Friends, belonging and emotional security are among the few things you can't find on Amazon, which is why our role of 'consumer' is consuming us.

This social engineering really took off in the early twentieth century, largely led by the nephew of Sigmund Freud, Edward Bernays. Bernays saw the general public as irrational and subject to herd instinct, and he began to use his knowledge of psychology and psychoanalysis to control social behaviour. One of his most famous campaigns was driven by the tobacco industry, and was known as 'Torches of Freedom'.

Before the twentieth century, it was seen as corrupt and unfeminine for women to smoke. When first-wave feminism took off in the early 1900s, Bernays thought he could exploit women's aspirations for freedom and equality by presenting cigarettes as symbols of emancipation. He said that 'Cigarettes were a symbol of the penis and of male sexual power . . . Women would smoke because it was then that they'd have their own penises.' During a large feminism parade in 1929, he hired women to disperse themselves among the crowd and march while smoking their 'torches of freedom'. His own photographers recorded the parade, and when the footage was released, the campaign sparked discussion around the world. The targeting of women by tobacco advertising led to higher rates of smoking among them: 'In 1923 women only purchased 5 per cent of cigarettes sold, in 1929 that percentage increased to 12 per cent, in 1935 to 18.1 per cent, peaking in 1965 at 33.3 cent, and remaining at this level until 1977.' Even today, a subconscious stereotype exists that means we tend to conflate female smokers with independence and nonconformity. This shows just how easy it is for companies to commodify our desires and for products to represent ideals and ways of life, locking us into an 'iron cage' of consumerism.

So what does all this have to do with plastic?

Well, in a society where our mantra seems to be 'I shop, therefore I am', and where our self-worth and identity are in a messy embrace with our purchasing habits, the more we buy, the more waste we produce. We have produced 9.2 billion tonnes of plastic in our time on earth, of which 7 billion tonnes is already waste polluting the planet. However, 'polluting the planet' is another misleading term, because there are pools of pollution concentrated in very specific areas, whilst other places remain almost pollution-free. Countries that have developed (but not perfected) waste-management systems often ship their own influx of waste over to developing countries such as Malaysia, India, Thailand and Vietnam. During the filming in India of *ANIMAL*, a documentary I worked on about finding solutions to the sixth mass extinction, we met people barely subsisting, working eighty to ninety hours a week so that the developed world can continue to suffocate under piles of unnecessary plastic waste (which contributes to our own brand of misery). For many, it has been a battle against the literal weight of such a powerful, permeating force, but individuals across the world are continuing the fight from their own corners of the globe, reluctant to allow the plastic tide to consume their communities.

'Welcome to Bali! Do you have any plastic bags to declare?' is probably the first thing you'll be asked when you arrive on the island of Bali. It all began with two sisters, Melati and Isabel, who were just ten and twelve at the time. They grew up in Bali, but Melati paints a different picture to the terraced rice fields and white beaches that might spring into your mind. During her childhood, she had seen plastic everywhere – in the river by her house, on the beaches, in the streets leading to school, and every year during the annual 'waste season', when the ocean sweeps rubbish onto the land as winter brings snow, or spring brings new life. The government dismissed it as a 'natural phenomenon', and so it took Melati until the day she saw a landfill near her home to piece the fragments together. 'There were cows grazing on the plastic, birds circling above and then diving into the rubbish to find hidden treasures. There were people too, scavenging through, and it simply broke my heart.' To Melati, nothing seemed natural about the human waste suffocating the landscape.

How did Bali transform from the island that dismissed the tsunami of plastic as a 'natural phenomenon' to a place that confiscates plastic from people at the airport?

In 2015, Melati and Isabel took a petition containing

100,000 signatures to the governor of Bali asking him to ban plastic bags on the island. They were turned away at the door. When they went home, they drew inspiration from their hero, Mahatma Gandhi, and began a hunger strike, unsure of how long they would need to persist before their voices were heard. After forty-eight hours, the governor agreed to meet them, and with the two girls watching on, he signed legislation promising to ban all plastic bags on the island by 2018.

Around the same time, 4,000 miles away, someone else was fighting the same battle.

I want to take you to Mumbai, in Maharashtra, India, to meet Afroz Shah. The chaos of arriving in India reminded me of arriving in a rainforest, except people and vehicles replaced the monkeys and birds. It was thrumming with life, a roiling tumult of mayhem and people, people everywhere! It's unspeakably exciting to stand on a street corner and have rickshaws, people and tuk-tuks weaving around you in a blur. Everyone is just living. Although you can barely move and breathe in the ebb and flow of traffic and people, the world feels invariably bigger, and you invariably smaller. But even in a city that seems to expand around you, one man has left an almighty mark.

Afroz Shah is a young Indian lawyer from Mumbai, who orchestrated the world's largest beach clean-up project. In 2015, with his neighbour, eighty-four-year-old Harbansh Mathur, he decided to tackle the mountain of rubbish that had accumulated on Versova Beach. Painstakingly they removed every bit of it piece by piece, working every weekend. The clean-up quickly grew into a global movement, with many groups around the world following Afroz's lead.

Afroz explained, 'Being a lawyer, your first instinct is usually to go and complain. That would have been an easy journey for me. But the responsibility was mine. I started to look at how I had helped to destroy the oceans and how I could rectify that . . . so I started cleaning. It began with Versova Beach, where I had grown up. I sat down with my eighty-four-year-old neighbour, sitting by the window overlooking what was one of the most polluted beaches in the world, and I said to him, "I'm going to clean this beach. Will you join me?" He looked at me like a bubbly teenager and said, "Yes, Afroz, I'll come with you."'

In the first week of October, Afroz and Harbansh walked down to the beach armed with a pair of gloves and some bags. They picked up five bags of plastic. 'That first day was uncomfortable. My hand would slip into the mountain of rubbish and go straight into somebody's shit.' Afroz smiles sadly. 'However, those circumstances made us bond.'

Within three years they had gathered a community and picked up more than five million kilograms of rubbish. Then, on a muggy morning in 2018, as the waves boomed and pounded against the sand, something happened that hadn't happened in over two decades – the hatching of 80 Olive Ridley turtles. Eighty little faces punctured 80 eggs, and 160 flippers scraped and scrambled towards the sea. It was a small moment, but it was symbolic of so much more.

After Versova Beach, Afroz began clearing plastic from

a local police station and forest, then, along with his team, decided to tackle the Mithi River, which meanders for eighteen kilometres through the heart of Mumbai. Two million people live on the banks of this river, which is piled high with rubbish, much of which is swept into the ocean.

On 22 November 2019, I found myself on the bank of the Mithi River in Dharavi, one of the largest slums in Asia. A slum is defined as 'a squalid and overcrowded urban street or district inhabited by very poor people'. Many of us have a specific image of slums in our minds, and journalists also portray them this way. This is because there is public hunger for proof that there are others out there worse off than us, so that we feel better about our own circumstances, and often, so that we can sympathize rather than empathize. This definition, however, is at odds with the vibrancy, the life and the community of these slums. They are often the stitches in the fabric of a city, holding it together at the seams. Dharavi is not a hub of despair at all. It is not sentimentalization of squalor to recognize how rich a place it is. Not materially, but socially. In our society, which functions like clockwork on the surface but contains so many unhappy people, who am I to judge the society of others from my first impressions?

I want to introduce you to Sushil. Lack of technology means I can't speak with him to fill in the gaps in his story, but I can tell you what I know from our snatched conversations by that sluggish river.

The film crew and I were picking our way along a canal that was swollen with the weight of plastic, engorged like a freshly fed snake. Women were bent double sorting through the waste, infants slung across muscle-roped shoulders. Children shrieked with laughter as they played, hopping across thick blue pipes that served as walkways over the putrid water. When they saw us, they became curious and gathered around. Admittedly, I haven't been a true stranger very much in my life. Often the people I talk to are those I've known since I was very young, those I have some preconceived idea about because I've seen their social media feed, or listened to an interview with them, or been told about them by someone else. So how rare is it to stumble across another group of humans and just interact, not because you want to buy a coffee or a train ticket or ask for directions, but just because you can? The language difference was a pathway rather than a barrier. With the lack of words between us, we played hand games instead. They taught us clapping games and we made strange faces at each other, for no other reason except that it was amusing, to us and them. Then, from the large group, they began to push forward individuals who could speak a few words of English. One of them was Sushil.

The children balanced themselves on the wall between us and the river. We sat down with Sushil and listened.

He had grown up in a poverty hotspot, in an informal settlement unregulated by the government. When we visited his home, people were drinking from the river, which was a heaving, viscous soup of plastic. It was toxic and putrid, and it was also where they bathed and played. Much of this waste wasn't theirs. In the rainy season, the engorged rivers sweep the city's waste along and it ends up accumulating near the mouth of the river, in slums such as this one. During the monsoon season, the grimy water also enters homes, so families have to cope with the sludge, waste and stench. So this plastic hadn't been accumulating for years, as I had assumed. It was all fresh. Only a few months ago, the water had swept the old plastic out to sea and this was just the newest wave.

When we left this area, we received a video of Sushil, with many other local kids, throwing the plastic on a big bonfire because they said we'd opened their eyes to what was happening. There were toxic black clouds rising from the fire. In their attempts to cleanse their homes, they were unknowingly poisoning themselves. A paradox of injustice.

Sushil had also shown us a blurry video of the raging river just a few weeks earlier purging itself of the plastic, only to be filled up again, and had talked about his community's feelings of government negligence. The prime minister, Narendra Modi, had pledged to clean up the country by banning plastic bags, but Sushil said that the enforcement of the ban is so hilariously non-existent that many people don't even know about it. This is similar to 'conspicuous conservation', which is when consumers purchase sustainable products in order to signal a higher social status, or to boost their social status. Many politicians now overemphasize their green credentials in order to gain votes with the booming number of people who value sustainability and environmental protection.

This conspicuous conservation, or 'greenwashing', is prevalent in advertising too. For example, in June 2018,

McDonald's, who used eight million plastic straws every day in the UK alone, announced that they would be swapping their single-use plastic straws for a paper alternative. I remember the joy on Twitter as people thought something was shifting, as though somehow this corporate giant had had an ecological epiphany. Then a leaked internal memo, reported on by the *Telegraph*, revealed that the new straws weren't recyclable after all, and that McDonald's were incinerating them rather than recycling them.

Modi's announcement about banning plastic bags is to environmental action what McDonald's is to food – a synthetic, detrimental version of the real thing.

Sometimes, though, as the government's capabilities diminish, citizens expand to fill that space.

Afroz and his team do this almost every morning. The volunteers pick up the waste, then tirelessly pass by each house to collect plastic and educate the occupants. Once collected, these plastics go to factories to be recycled. A good part of the time, the work of sorting and degrading plastic is done in slums such as Dharavi. After seeing women spend twelve hours on the ground sorting bottles that have only been used once or twice, you can't help but think that recycling should only be done when there's no other option.

Even for people who do care enough to recycle, what goes on behind the scenes is often shocking.

The problem first emerged when we began to recycle for the planet rather than ourselves. After the Industrial Revolution, there wasn't such a great need for people to recycle goods for economic purposes. When products were so cheap and quick to produce, it made more economic sense to throw old things away and just buy new ones.

Recycling is almost an intrinsic human trait. There are records of Athenians in 500 BC organizing the first municipal dump programme in the Western world. The citizens had to dispose of their waste at least one mile from the city walls (and we think we have it tough today having to haul it out onto the pavement). Yet despite its long history, recycling has still failed to stave off environmental and social catastrophe. The truth is that our current system is not as effective as it seems, for two main reasons:

- It helps us dodge the responsibility for our rampant and unsustainable consumption.
- Lots of our 'recycling' is never actually repurposed. The remnants of our waste that can't be recycled because they are not clean enough or are made of the wrong material are called 'residual'. Much of the waste we recycle ends up as residual in landfill because it doesn't meet the standards set.

Up until early 2018, much of the world's waste was shipped to China, but in a move that shook the waste industry, the Chinese government decided to adopt more severe policies, meaning they essentially banned the import of foreign waste. We were in chaos. Countries like the UK, the US and Germany began scrambling around in a race to find other South East Asian countries to ship waste to.

In these new locations the separation of rubbish is done primarily using low-cost labour. The workers are consistently exposed to toxic chemicals for minimal pay and in terrible conditions. The injustices of waste's impact on the planet are often considered, but we rarely factor into the equation the cost of the exploitative labour system that powers the global waste and recycling system. Pure good and evil rarely exist in the environmental movement. It is always immeasurably more complex than it seems.

Recycling does reduce waste, save energy in manufacturing and create jobs (some 1.25 million in the United States alone), but the fact is that it is not a powerful enough tool to create sufficient change on its own.

Afroz has mobilized residents, Bollywood stars, schoolchildren and elected officials, and he has also been knocking on the doors of the fishermen. Together they have collected nearly nine tonnes of plastic in three years. Versova Beach has been completely cleaned. However beautiful it was to see their active passion and effort, though, it wasn't enough on its own. Together we cleaned up the forest, the beach and the river, then peeled off our gloves and proudly surveyed our good work, but then just a few days later more people donned new gloves and the whole process started again. The same thing happened in Bali. Melati worked to get plastic banned on the island, but even a complete ban didn't stop the long crescents of golden beaches being contaminated by heaps of waste and littered with washed-up remains of Olive Ridley turtles and Bryde's whales that had died after ingesting plastic waste.

It is like trying to stop a bath from overflowing without turning off the tap – as well as cleaning up the existing mess, we need to stop the plastic at its source. The reason is obvious: infinite growth of anything in a finite system is unsustainable and fatal. In the human body we call it cancer, yet on earth we call it progress.

Here are some of the dominant narratives about 'progress' meticulously crafted and pushed by the plastics industry:

- plastic pollution is an oceans problem (subtext: we need to focus our efforts on the clean-up rather than actually stopping the production of plastic in the first place)
- the global south, places such as India, is the main culprit in plastic pollution (subtext: companies from the global north are able to keep pushing more plastic production)
- recycling is the solution (subtext: the fault lies primarily with the consumer, so the big corporations producing the plastic are exempt from blame)

On our first day with Afroz, we went to Sanjay Gandhi National Park to do a litter pick with him and the local community. We worked with them labouring in rotting rubbish under the scorching Indian sun. To pass the time, the group began to share their stories, digging into their deep wells of experience. It reminded me that I was only doing this for a few days and would then return to the manicured lawns and spotless streets of the UK, while these people would spend their lives fighting what felt like a losing battle.

One of them was Nandil, who was seventeen at the time. He, like Afroz, was driven by the dream of what could be achieved if everyone took action. He was doggedly determined to play his part. When I asked him where his journey had started, he recounted an experience just a few months before. He had entered the park with his friends and seen the forest wardens surrounding a dead deer – a chital. She was golden, with splashes of pure white across her muscled velvet body. Beautiful, young and healthy, until they examined the body and twenty kilos of plastic spilled out, along with her unborn baby, entangled in the mess of humanity before it had even taken its first breath.

The thing is, plastic is not just an ocean problem, and so there is not just an ocean clean-up solution as the dominant narrative tells us. Instead, we must clamber down the supply chain and examine the root, which, as we've discussed, is our consumption. Often, breaking something down is a precursor to it growing back stronger and more resilient. Before a caterpillar becomes a butterfly, it becomes trapped in the chrysalis and loses all structure as it becomes a gooey liquid. However, the cells within this liquid then have the opportunity to organize and re-form themselves into a butterfly. Right now, we are still like caterpillars. Everything we know and rely on has become fluid and unstable as scientists predict a descent into chaos. Our normal is being shattered. This is the time to rebuild, but rebuild stronger.

This faith in such a huge societal change isn't a blind hope that ignores the magnitude of our crisis, but a practical one that sees all the ugly facts and decides that we are not going to face the end of the world, but only the end of the world as we know it.

So how can we break free from this cycle of consumption and relentless amassing of superfluous things that we're currently trapped in, and rebuild something better? We can find hope in another type of cycle: the circular economy.

When you hear the words 'circular economy' ('designing out waste and pollution, keeping products and materials in use, and regenerating natural systems'), your mind may immediately leap to recycling as a means to that end. Recycling is a false friend, though! In a true circular economy, the focus should be directed towards avoiding the recycling stage, and then only using it when necessary. Preventing the waste from being created in the first place is the most effective strategy. While recycling must be used as we transition towards a circular model, putting the emphasis on genuine circular innovations will move us away from our current waste-based model. Recycling only begins at the disposal stage, which means that products are not designed in a way that will increase their value, enhance their materials or preserve the product itself.

Waste, at the moment, is an economic externality. This just means that producers see it as someone else's responsibility. In the current system, that manifests as 'planned obsolescence' – in other words, designing a product to break so the consumer must keep coming back for more and more. The idea of planned obsolescence has been floating around for a while, yet never seems to break through the barrier that separates subliminal and mainstream awareness. The film *The Man in the White Suit*, made in 1951, follows a chemist who creates a material that shuns the idea of planned obsolescence because it never needs replacing. He is soon attacked by textile producers and trade unions. Many decades on, we are still drenched in the detritus of planned obsolescence. The *Guardian* reports that 'the average lifetime of desktop printers is a mere five hours and four minutes of actual printing time, and an individual keeps a smartphone for between two to three years usually'.

This current system (linear/reuse) can be summarized as 'take, make, dispose'. For a company to operate successfully in the linear economy, they will try to buy material for the cheapest price possible and then sell as many manufactured products as they can without thinking about the waste this creates. This way of working assumes materials are infinite. A company in the circular economy would change this 'take, make, dispose' into 'make, use, return'. Their focus would be on extending the life of the product, reconditioning activities, and waste prevention.

Let's use a light-bulb company as an example. If the company actually own the light bulbs, this gives them the incentive to make sure the bulbs they produce are energy-efficient and resiliently built. People would then lease the bulbs, and return them for maintenance and replacement. This radically different business model turns companies into service providers rather than sellers, which is actually beneficial for the companies, because they can hold on to customers for years. For example, if the bulb fused, rather than buying a new one from some other company, the customer would get it repaired or replaced by the same company, as they are already paying a subscription for it.

This circularity is a value system I heard mentioned a lot as I spoke to young people from different corners of the world. One of them, Maya Chipana, whose family are Native American from Bolivia, told me that for indigenous people there is a natural acceptance that very little in nature is linear. To them, there are many unseen forces at play, inputs and outputs and links that must be considered. Energy conservation is not a new buzz word; it is an ancient belief because they understand the interconnectedness of humans with nature and our place in the world.

Maya grew up in New York, but her family is from Altiplano, a windswept plateau that stretches between the rugged Andes mountains. A coloured vista peppered with flocks of pink flamingos, flecked with brick-red lakes and the azure waters of Lake Titicaca. On the Zoom call, her younger siblings race around in the background, scrunching up their faces and sticking their tongues out at the camera. Maya tells me about the Keetoowah people, her ancestors. Her parents, both indigenous, from rural backgrounds, crossed paths in New York, so Maya had a city childhood yet still managed to avoid gaining the narrow vision that a blinkered urban existence and lifestyle imprint on us, with the city's thick, confining slabs of concrete, tarmac and glass. When she goes to visit her grandparents and family in their remote village in Bolivia, she is struck by how humble and elemental it is.

She tells me that there is 'no Wi-Fi, no fridges, and no discontinuity between man and nature'. They see the trees as relatives: 'we co-exist and we provide them with CO_2 whilst they give us oxygen. We lose that in an industrial society, it makes us pollute without realizing that we're also polluting ourselves.

'Most houses still have adobe or dirt floors, so before meals you pour a bit of your beverage onto the floor to recognize that you are giving back, whilst taking food. As Native Americans, each time we eat we also prepare another plate of food; a "spirit plate", which we put out. It does not matter whether it is consumed by the hungry, the homeless, the birds, dogs or cats, the important thing is recognizing that we share, and we give back.'

Why is this relevant to the plastic crisis? Because this is the mindset and course of action we need to adopt, one that recognizes that absolutely nothing is linear. I challenge you to come up with one thing, one concept, one process in nature that does not at some point circle back around. The average modern adult likes straight lines: logical, convergent thinking and clear, linear career paths. Children, however, find it harder to think, speak or move in a straight line. It's deeply unnatural. They'd rather bounce loosely down the road, skipping this way and that. They pull their sentences in different ways, leaping from idea to idea, pouncing on every fanciful dream that occurs to them. Young people are not yet bound by the civility of straight lines, but are rather floating happily in their own non-linear existence. We all live in biological cycles that feed back into the system. Digestion, life and death, even sleep. However, we must realign our technological cycles. That means recovering and restoring materials and products by reusing, repairing, remanufacturing or (as a last resort) recycling.

There is still a long way to go before the plastic stops pouring into the oceans, but the world is making progress. When we were in India, Afroz spoke of how the government wants to implement a national plastic ban by 2022. On 24 October 2018, the European Parliament also voted to ban certain single-use plastic objects. Cotton buds, straws, plates, cutlery, etc. account for 50 per cent of plastic pollution in the oceans. The ban was due to come into force in 2021. It was achieved despite intense lobbying by the plastics industry. Amsterdam is embracing a 'doughnut' circular-economy model to mend the post-coronavirus economy. Things are changing.

The feeling that many of us have that there is more to life than consumerism is not wrong, but the way we act upon it is. Beyond consumerism lies a society of enough, and the way to get there is to not be swayed by extrinsic motivation (profit, success, status, fashion), but instead to tap into our intrinsic values (creativity, compassion and community). Imagine a future where we don't feel our value is based on material things. Mindset is the foundation, and change will come when that mindset translates into action. Here are some ways you can be a part of the change.

Manifesto for change

- Incentivize the design for circularity. Buy ethically, but also let companies know why you're doing this. This is the most effective way to drive change in companies who want to appeal to their customers.
- Discourage conspicuous consumption. It's part of human nature to want to conform, and whilst we can't change that desire, we *can* change what we conform to. This doesn't mean publicly shaming people who are buying new things, but instead helping to create a culture where we value possessions more deeply and see it as fundamentally flawed to dispose of things carelessly.
- Stop copying other people. Don't chase the latest trend. Don't become prey to the illusion that conformity is the key to happiness. Rather, conformity is the key to an endless series of doors that lead nowhere and leave you empty and tired from the pursuit.
- Look at the true cost of what you buy. Look beyond the price tag and think about the amount of time and energy you are exchanging in order to purchase something. You could have spent that with family or friends, or in nature. Consider whether a product is really worth it. Think about the cost to the other lives along the supply chain.

Were they treated ethically? Was the item produced sustainably?

- Participate in 'resale disruptors' – online consignment sites where you can buy goods second-hand and sell your own unwanted items. If you do buy something online, try to use sites like eBay, Depop, thredUP, or hewi. Studies have shown that resale is actually overtaking fast fashion now, with resale disruptors growing over twenty-four times faster than the retail industry.
- Create a community of collaborative consumption. Collaborative consumption is resale disrupting on a large scale. It is also known as the sharing economy, and you've probably participated in it if you've ever been to flea markets, garage sales, car boot sales or charity shops, or participated in ride sharing, apartment sharing, house renting or couch surfing.

Chapter 2

The Air We Breathe

Air pollution, global travel and reducing emissions

According to the World Health Organization, air pollution kills seven million people every year. Traffic accidents, diabetes and AIDS combined have a lower death rate than that. We wear seat belts and spend a long time educating kids about road safety to prevent traffic accidents, we exercise and try to eat healthily to prevent diabetes, so why are we not willing to transition to greener energy or to consume less energy to prevent seven million deaths from air pollution and many more from the consequences of climate change? Among the likely impacts of repeated air poisoning are a 'huge reduction' in intelligence, stunted growth of lungs, dementia, mental health disorders, asthma, cancer, stroke, heart failure . . . The list goes on.

As climate change is a cumulative and relatively slow process, we often see it as abstract and distant. A study published in 2015 in the journal *Nature* by psychologist Adrian Brügger, titled 'Psychological responses to the proximity of climate change', found that 'Britons who had recently experienced flooding (expected to occur more frequently in Britain because of climate change) perceived their local area to be more at risk from climate change, were more concerned about climate change impacts, had higher confidence in their ability to mitigate climate change, and were more willing to reduce their energy use in order to mitigate climate change than those who had not recently experienced flooding.' It is this psychological distance that is part of the reason for our apathy. Although many of us have grown up with the sun hidden behind a pall of smoke, or unaware of the star-studded sky above, we still see air pollution as someone else's problem.

It's hard to identify the precise moment when the air pollution crisis moved from the horizon of popular imagination to our immediate reality. For me, it was July 2019 when I realized that the time of looming had ended and now was the time of consequences. I was doing an event with Change.org at the Tate Modern in London. I arrived early, venturing through pulsing light exhibitions and up endless winding marble staircases. The event was an interactive round table of campaigners debating with the public on topical social and environmental issues. Even though I had been an activist for several years by then, I was still painfully shy. I was always reassuring other young people of the power of their voice, but I still felt like an intruder in that space. A space that is overwhelmingly middle class, middle aged, male and white. Of course, we need everyone. But with a population so diverse and different, the environmental spaces shouldn't be so predictably similar, so fiercely exclusive. Perhaps it was my youthful self-consciousness, but that day I felt the immense weight of the words I did or did not say.

When the event began, I slipped quietly into my seat, waiting for my turn to speak. Next to me sat my polar opposite: Rosamund Kissi-Debrah, an outspoken activist, but rapturously so. She had stories to tell, jokes to deliver and rebuttals to fire. As she was so vivacious, so joyful, I was hit even harder when she told her story.

Rosamund was, and still is, vigorously campaigning for London's toxic air to be cleaned. It all began in 2013. Well, it really began in 2004, when her daughter Ella was born: 'Ella was born healthy. She was an active and happy child – cycling, skateboarding, playing football and excelling at swimming. She dreamt of becoming a pilot. She was always incredibly helpful. Then when she was six, she developed asthma, and the frantic hospital visits began. Yes, we knew that she was breathing the pollution in, but you really do not know what damage it is doing . . . until it has happened. Ella suffered three years of seizures, resuscitations and hospital stays.'

One wintry day in February 2013, when Ella was nine years old, she and Rosamund both began their day as they usually would. They could hear the rumble of traffic, because they lived twenty-five metres from the South Circular Road in Lewisham, one of south-east London's busiest roads. But at that moment, amidst the hustle of life and childhood, it was merely ambient noise. The thunderous traffic gave out petrol fumes, Ella knew, but at nine, why should she worry about that? It all disappeared after a few seconds anyway. Rosamund made breakfast and Ella picked out her outfit for the school disco. But she never made it to the party: 'Just a few hours later she suffered coughing fits followed by several seizures. She was hospitalized . . . and we lost her.'

Amidst the shards of tragedy, most of us turn inwards, trying to protect ourselves from the harsh reality of the world. However, Rosamund knew that the beast that had cut Ella's life so short was still permeating playgrounds and terrorizing streets. She became the proverbial dragon-slayer against an invisible foe. 'I was so angry that people knew. That the government knew, but whenever they found out, they certainly didn't go out and tell the general public.' She pushed for an investigation and her suspicions were proved right. 'It was discovered that Ella's hospital admissions were linked with spikes of illegal levels of air pollution around our home near the South Circular Road. Her final hospital admission took place during one of the worst air pollution episodes in our local area. A renowned medical consultant studying our case found there was a "real prospect that without illegal levels of air pollution Ella would not have died".'

In the UK, air pollution causes around 40,000 premature deaths per year, and levels of nitrogen dioxide have been illegally high since 2010 in the vast majority of urban areas. The government have been taken to court three times over the illegal levels of air pollution and lost each time. Yet the authorities are still failing to act. 'What is more valuable than the life of a child?' Rosamund asks.

Rosamund knows the investigation will not bring Ella back, but because it concluded that her cause of death was 'acute respiratory failure as a result of a severe asthma attack', she is the first formally classified British person to die as a result of air pollution. This ruling officially makes air pollution a human rights issue, as it breaches the 'right to life' that we all legally possess.

So, as citizens, how can we defend this right to life?

How can we create a world where roads are there for people, not traffic? Where the melody of our cities is humans and birdsong rather than the continuous growl of cars? Where children breathe fresh air as they walk to school rather than pushing through fat clouds of thick exhaust?

The same year Ella was killed by noxious air pollution, Kevin J. Patel was attending sixth grade over 5,000 miles away in Los Angeles. He was sitting in class on a normal day when his chest began to contract, his heart began to pound and his breath became shallow and halting. He recalls the nurses fluttering around like wild, frantic birds, and a rapid swell of paramedics whisking him away to hospital. Kevin was diagnosed with arrhythmia, a condition in which the heart beats differently to its usual pattern. He didn't smoke or do drugs. He was twelve years old. He did, however, live in South Central Los Angeles.

Los Angeles was first written about in 1542, and one of its first descriptions named it the 'Bay of the Smokes' because of the hazy fog hanging over the region. This wasn't industrial smog – the internal combustion engine and factories were still centuries away; this pollution was generated by local villages and held in place by the mountains that shelter LA, trapping whatever pollutants rise from the ground. The unique topography of the city provides a fatal foundation, exacerbating the already deadly fumes generated by humans. It's a transient place; people are drawn by the dream of it, streaming into the city in their cars and barely setting foot outside of them. William

Faulkner, whilst screenwriting there in the middle of the twentieth century, called it 'the plastic asshole of the world'. It is a land of gated communities, bottled water and thick-wheeled cars. A land of insulation, protecting those who can afford it from the consequences of their actions.

In an attempt to channel his fear and anger into something tangible, Kevin began his own organization, OneUp Action, through which young people could set up youth climate commissions to get involved in the development of environmental legislation within government. Now young people throughout the States work directly within the government to suggest and write policies to help fight climate change. Very soon after this first initiative, young people from all over the world started to contact Kevin asking for advice and support, so he created Action Chapters: a 'youth-led, intergenerational, and intersectional community of activists fighting for a regenerative future'. He provides mentorship and resources to young innovators and activists worldwide. Rather than being dismayed and deterred by his personal struggle against the system, Kevin cast a ripple of change across the water and is now feeling the stirring of a powerful undercurrent as things shift and transform.

Since an environmentalist is 'a person who is concerned about protecting the environment', and you are reading this book, you are also an environmentalist. We environmentalists – you, me and others who are also concerned about our planet – may fall into a dangerous trap as we learn more about the problems we're creating.

Just as some of the most intelligent people make terrible teachers because they expect everyone to learn as fast as they do, an environmentalist should generally not revert to talk of global quantitative statistics and abstract, complex graphs. When we do that, we frame the scale of the crisis in such a theoretical, grand way that paradoxically, instead of making people more concerned, we enable the continuation of the very activities generating CO_2 and pollution.

We humans are not psychologically and biologically equipped with the tools to be able to comprehend the sheer scale of what we have done to this planet. And there is very little use in trying. On a global level, a single power plant, or even a city, is only making a negligible contribution to the world's emissions and air pollution. Anyone can find a way to exempt themselves from this large, complex web of condemnation. However, when we shift our gaze from global to local, from theoretical to emotional, from 400 parts per million to Ella Kissi-Debrah, it becomes much harder to detach ourselves. It goes from a not-in-my-backyard issue, to a very-much-on-my-doorstep one. Or, should I say, in my garage . . .

We live in the age of the automobile. Okay, we live in the age of many things, of Netflix and Kim Kardashian, TikTok and Twitter. But physically, the crust of our planet is being moulded into an automobile's paradise. The number of cars on earth is expected to rise from 1.2 billion to 2 billion by 2035.

Never have we been so mobile, so free to go from A and B with little to no thought, yet never have we been so obese, depressed and socially atomized. A large reason for this is transport, mainly cars, or the people in those cars. The coronavirus was like an MRI scan, exposing the inequalities, poor governance, corruption and systemic problems embedded in each place it touched, but it was also an ephemeral portal into a future where cities are less polluted and, in a strange way, more community based. When I cycled along the road during lockdown, I could actually feel the wind in my hair, rather than having to weave haltingly through traffic. There were children roller-skating along the road. Ducks and deer found themselves using the new-found space.

Yet whilst it was a hallucinatory, rosy few months for some, many were left sitting in the crumbled remains of the lives they had built. And that was the problem. The reduction in pollution was not significant enough to warrant the suffering the virus brought. It did, however, give us a taste of what the future could be, and the good parts of it were addictive.

Over time, we have seen our moral imagination stagnate and atrophy. We are enticed by the comfort of 'normal' because it is easier, but normality is a concept that limits our imaginations. A 'normal' world is like a 'normal' human. If we are constantly trying to conform, the unique individual self disappears; it takes a lot of energy and loss of individual creativity to fit in, often resulting in psychological problems. We feel more comfortable, though, when we're not deemed weird by society's standards, and so we begin to believe that this is the safest, easiest option. The same goes for our planet and society as a whole. Where we are now *feels* safe, but at some point our society will have a breakdown. It just won't be confined to the psyche in this case.

So before this happens, let's take a second to imagine how we might go about building a future without mass suffering, insecurity and unemployment.

The use of fossil fuels is woven into every facet of modernity, from money and medicine to manufacturing, housing, agriculture and transport. Many activists demand an end to fossil fuels, which in itself, as a demand, is useless. There is no one person, country or organizing body that can single-handedly remove fossil fuels from the equation.

There are two main reasons why air pollution has become as invisible in the global dialogue as it is to the naked eye.

- Many philanthropists and organizations have mobilized against polio, tuberculosis and malaria. There is no massive global lobby trying to spread these diseases. If you take on these issues, you are heroic to everyone. If you take on the issue of fossil fuels and pollution, you have to fight against commercial interests and almighty industries who want to silence you so that they can continue making vast amounts of money.
- To tackle the diseases mentioned above, you can take discrete and measurable actions, like distributing mosquito nets or funding

> vaccinations. To fight air pollution, you cannot just hand out face masks and say the job is done. If you really want to address the issue, you must intervene systemically, which requires a great awareness of the various contributing aspects: the economy, human behaviour, etc. Pollution is a by-product of corruption, and to take on pollution is to take on that corruption.

In this the age of the automobile, we live in a vehicle's paradise rather than our own. This is CARmageddon. Reducing our use of cars is the key to much more than reducing pollution. It will create stronger communities, a better social life, more green space and nature. In the future, the efficiency of the products and objects we use will be key. So how would we achieve this in terms of transport? We would, as expected, turn to bikes and more efficient modes of public transport. For those who require cars, for occasional trips out of the city, you would be able to hire one or join a car club. The average car sits outside, idle and unused, for 96 per cent of its life. There has to be a more efficient way to provide for this average of seven hours a week when you want it.

Right now, a future in which everyone travels in driverless flying cars still dominates the popular imagination, but what if the best future was not one of more, but instead of less? Imagine a world with fewer cars in cities. Imagine children playing on major urban thoroughfares dense with wildflower meadows and thrumming with bees and butterflies. Cafés and restaurants spilling out into the road. Pedestrians walking nonchalantly down the middle of the street. Whilst it may seem like a restriction of our fundamental freedom to remove private cars from city centres, in many ways *not* limiting cars is the greatest restriction of freedom, of childhood and of opportunities for future generations. But is this possible?

It has been done. Many times, in many different places. La Cumbrecita in Argentina has become completely pedestrianized, with no transport at all. Venice, Italy, uses only boats. Halibut Cove in Alaska requires skis. Then there's Helsinki, in Finland, where they are creating a revolutionary system that has somehow escaped the world's notice. Their mobility-on-demand scheme is described as a way to 'transcend conventional public transport . . . The hope is to furnish riders with an array of options so cheap, flexible, and well coordinated that it becomes competitive with private car ownership not merely on cost, but on convenience and ease of use. Helsinki residents will use a new app to simply indicate start and end points, with perhaps a few preferences for mode of transit. The app would then function as both a journey planner and universal payment platform, knitting everything from driverless cars and nimble little buses to shared bikes and ferries into a single, supple mesh of mobility. The app would be like Google Maps mated with a public Uber, but across all transportation options.'

So once again, we're touching upon something profound. Some believe that as individuals, our purpose is not to come into this world to make it better. Some believe that not every one of us should have to 'sacrifice' cars, fossil fuels and single-use plastic. I would agree that the whole aim and structure of our lives cannot just be about leaving the planet better than we found it. For if we are always thinking about the future, who will ever think about the present, and find joy in the beauty that has been preserved by the ancestors who did think about saving the world? No one can save the planet in a single lifetime; we can't even save ourselves from our inevitable fate! However, if you want to build your life around joy, love and amusement, the path to saving the planet, by default, usually fulfils these criteria.

Studies have shown that children living close to busy roads have a higher risk of hyperactivity and emotional problems. There is less social cohesion in the community, and people report higher levels of unhappiness and discontentment with their life. The health of humanity is so inextricably bound up with the health of the environment that in most cases, every step we take to protect the natural world has a domino effect. The advantages multiply and ameliorate with irresistible momentum. By protecting nature, we too will be able to live in a reality of our own making.

If your main resistance to minimizing the number of cars on the road is freedom and choice, surely there should be a hierarchy of these privileges. The freedom of choosing a long life over a premature death should take precedence over whether we cycle or drive. Freedom should mean our brains and lungs are not brimming with metal particles and our mental development is not stunted because of the road we happen to live on. So if your freedom is stewing in a traffic jam in central London, perhaps you need to redefine what 'freedom' means.

Of course, the most important thing is that we have a diversity of solutions. Completely ridding roads of vehicles is not going to happen, and so alongside reducing vehicle use, there's also the pathway of changing the type of vehicle we use, or the kind of energy we use to power it. Stricter government regulations and reducing personal consumption do not have to be at odds with innovation and new technology. In fact, some of the young people I know who are leading the way in battling pollution are using innovative technologies as their weapons.

One of them is Param Jaggi, who began changing the world for the better at the age of twelve. Before that, he had been a watchful child, the curious one. The child who would deconstruct things until their essence was laid out in front of him, always willing to surrender ideas and absorb new ones based on his findings. But for Param, it was something more, an inherent part of himself that he clung onto during the turbulent years when most people let it dissipate. His is a classic tale of ingenuity and passion, of a boy with a heart full of volatile crazy dreams that would fizz and splutter until they could no longer be contained.

One of these dreams was born in his parents' car, at a red stop sign. As it ground to a halt amidst a throng of other cars, Param, barely big enough to see out of the windscreen, noticed how the vehicles' emissions were pooling around them in a noxious grey cloud. The easy logic of a young brain made him realize that if these fumes he could see kept coming out of the almost two billion cars around the world at the rate they were now, the atmosphere would be full of them by the time he himself was old enough to drive. He returned home concerned, and this one revelation turned him from Lego to lab.

He recalls splitting his bedroom in two and dedicating half of it to what was to be his greatest invention yet. It was from a frenzy of duct tape and metal – brought to life off a napkin sketch – that the first prototype of the Algae Mobile was created. It is a device that converts carbon dioxide from cars into oxygen. He gained a patent for it, and since 2009 has been building and rebuilding improved models. He won an award from the US Environmental Protection Agency, and then went on to set up his own company, EcoViate, which is now a leader in green and alternative energy. In the same way that Param began by tearing things apart to learn a lesson he could later use to create and invent, we too as a species must learn lessons from the destruction and tearing apart that we have done and mature into humans who use those lessons to build something new, something better.

One of the best things about new forms of energy as a result of innovation, whether geothermal, wind, solar or any other, is that it allows for decentralization. Where before many of us were reliant on a single expensive energy grid driven by fossil fuels, now individuals are installing their own solar panels or using alternative sources, such as biomass. Although it has until now been a passion project for those in already industrialized countries, communities in lower-income countries are skipping the weighty process of fossil-fuel-driven centralized energy and installing their own solar panels to generate the energy they need. Changing your car to electric (when the time comes), changing your home energy supplier and persuading businesses (who are at the vanguard of the energy transition movement) to invest in renewable alternatives is essential.

There's one other thing we should discuss, a passion of mine that I've had to accept I won't always be able to fulfil: adventure and the exploration of the unknown. My brother is one of those people who confuses me greatly. I'm comfortable in and seek out new and volatile environments. He's comfortable with conformity and consistency. His idea of exotic is a trip to the Tesco just down the road. The biting zeal and need for change that makes me so restless may have a genetic root. A study in 2015 highlighted a gene related to the desire to wander. They called it the 'wanderlust gene', as it produces a greater likelihood of novelty-seeking behaviour, impulsivity and adventurousness. I like the idea that you can pack everything up and leave the social sphere of your house, city or country, to be enveloped into a larger world where you're free to be anyone. I love the idea that however old you are and however far you move, there will always be stories playing out that will never quite intersect with yours.

My theory is that humans make their environments as consistent and comfortable as possible so that the effort they have to exert daily is minimal. However, those cheesy Instagram quotes such as 'travel broadens the mind' and 'life is short and the world is wide' actually have a very valid point. Despite much of our motivation behind travelling being the chance to plaster our Instagram grid with turquoise seas and martinis, there is also the aspect of wanderlust. It's strongly embedded in human evolution. We lived in nomadic communities for almost 99 per cent of our history, following the wind, quite literally. We went where the seasons and hunting patterns led us, and walked where our hearts desired. Around 10,000 years ago, we began to fence and enclose plants and animals, and became fenced and enclosed ourselves. But within our increasingly stationary bodies, we still had the same wandering minds seeking novelty.

When you think about exploration and novelty, travel makes sense, but too often now, travel is about relaxation, not exploration. The fact that some people are willing to pay thousands of pounds and travel thousands of miles to lie in the exact same position under the exact same sun as they would in their own garden seems a little strange. As with consumption, the way in which many travel now is like a pursuit with no purpose. We want to untether ourselves from the person we are in ordinary life, but often realize when we arrive on our holiday that we are still glued to social media, our mosquito bites are unbearably itchy, and after all that effort of travelling to get away from ourselves, we have inadvertently brought the reality we are trying to escape from with us.

Most of us were stumbling around in the dark at the beginning of the pandemic trying to occupy ourselves, but then, like bursting into the light of a forest clearing, many discovered passions and hobbies we'd never even known we had. As confinement began to lift, I cycled through the swarm of fat flies that hang by the river on heady summer evenings. My wheels bumped over fallen plums, their purple-black flesh pushing out of their open skins. I put my bike down by the river's edge and lay on the pale, deadening grass watching families kayaking and swimming in the Thames. Young children were queuing up by the faded ice-cream van, an old couple were holding hands on a bench. It was a typical summer's evening, and I felt like I was on holiday. I had walked that path many times with a rucksack of school books on my way home the previous summer, and had been annoyed by the sweltering evening heat and the people taking up the path with their inflatables. It had not felt like a holiday back then. The only difference now was the new mindset I'd been in.

That's where the 'staycation' comes in. It is not a new concept. In fact, for many people, up until very recently, the staycation has been the only attainable means of having a holiday. To a young child, sand is sand, ocean is ocean, and Bournemouth Beach could well be in the Maldives for all they care. At what age do we start seeing the landscape as something to be crossed to get somewhere else rather than something to be explored and uncovered in its own right? I did go abroad when I was a young child, and I remember snippets of the startlingly bright, garish plastic slides and the all-you-can-eat buffets. I remember the warm ocean and walking along terracotta tiles to get back to our hotel room. But I also remember going to Dorset on the British coast. Those haphazard holidays with their unpredictable weather and ineffable perfection are no less cherished in my memory than those where we took a plane. In a way, there is something unparalleled and satisfying about the staycation, and I think it will play a larger and larger role in how we use our holidays in the future.

Perhaps this craving for something foreign is really just a craving for something wild, something natural. Thirty years ago, flying over Knepp Castle in West Sussex, you would have seen the conventional gridlocked prison of yellow and green squares bordered by uniform dark-green lines that characterizes much of the British countryside. No wonder we're sick of it! The same fields that are cultivated to feed flocks of sheep are expected to fulfil our inherent and intense biophilic needs. Now the view of Knepp feels foreign in its scraggy ferocity: blotches of scrub interspersed, like the African bush, with unrestrained free-moving herds of grazing animals. The area has been rewilded and now resembles what much of the British countryside would have looked like without intensive agriculture and overexploitation.

But for those times when we must travel abroad, well . . .

Flying around the world for business meetings several times a year is something we must end for obvious reasons, and with obvious solutions. I can imagine a world in 2050 where we laugh about the fact that that ever happened. Videoconferencing could save US and UK firms a total of $19 billion, as well as reducing CO_2 emissions by 5.5 million tonnes.

Travelling for holidays is more complex, because the paradox of not being able to see the planet you are trying to protect isn't lost on me. Hurry, says a voice in my head, go and explore before everything has been discovered, go and see species before they're gone! No, shouts back another voice, because doing so accelerates their disappearance.

So now we're plunged into this sticky moral crisis where we must ask ourselves if our personal enrichment and privilege is enough to justify endangering other human and animal lives. We don't have the right to be everywhere. My love of seeing the world is not wrong – it is beautiful to want to explore and discover – but it's arrogant and entitled to feel that I must fly when most people have, until recently, lived in one part of the world, and stayed there. This entitlement is symptomatic of the larger problem, which is that for the most part, the people who can fly aren't being affected by the first wave of the climate crisis. Privilege is like a private jet parked in the driveway, providing reassurance that when the wave hits, you can fly yourself to the top of a pristine hill and look down on the quagmire of sludge (although that is an illusion, because even that hill will not be exempt from the effects).

To travel without flying means using ground-based methods, also known as 'slow travel'. Apart from taking trains through Europe on a few occasions, I'm still largely a stranger to slow travel. I wanted to speak to one of my friends who had done it, and on a huge scale too . . .

Tori Tsui grew up in Hong Kong, a place famously known for being a victim of Chinese pollution. She says she remembers as a young girl trying to engage in conversations about waste, air pollution and the effects they were having on people and the environment. Then, in 2019, after attending university and now living in the UK, she attended an Extinction Rebellion protest, where she was scouted by the well-known sustainable fashion designer Stella McCartney, who asked her to feature in a campaign for her brand. Tori and three other activists, dressed in earthy colours, were snapped wearing McCartney's clothes and plastered on her Instagram. The comments were a mix of scathing criticism pointing out the fact that Extinction Rebellion had recently proposed a year-long boycott on buying new clothes, and also unshakeable support.

Stella then proposed to Tori and thirty-five other young activists that she would sponsor them to sail across the Atlantic and attend the UN Climate Change Conference COP 25, all without setting foot on an aeroplane. The plan was to depart from Amsterdam on 2 October. They would be at sea for almost two months, reaching Rio de Janeiro around the end of November. The last part, overland to Chile, would be done by bus.

Also in 2019, on a rain-streaked August day, a crowd stood like sentinels on the edge of Manhattan watching dark sails materialize out of the ocean haze. On board the yacht *Malizia II*, a small figure swamped in an oversized black jacket, a single braid tumbling down her back, raised an arm and waved. A cheer rippled through the spectators as they watched Greta Thunberg and her team make halting progress towards the US shore, in an attempt to attend COP 25 without taking a plane, after weeks at sea crossing the Atlantic.

By rejecting the conventional method, both Tori and Greta sent a message to the world about the dangers of how we currently travel. However, a sixty-foot racing yacht with solar panels and underwater turbines making it carbon neutral is not a means of transport just anyone can choose. So rather than interpret these journeys as examples we must follow exactly, they are there as guidelines, symbolic of the fact that there is a need to find new ways of doing things.

The message Greta was trying to convey did work. It led many to eschew air travel altogether, and the term *flygskam* ('flight shame' in English) took off in Europe. A Bloomberg report showed that the number of people flying between German cities dropped 12 per cent in November 2019 compared with a year earlier. There is not one single new way of doing things; rather we each must look at our own life and adapt it accordingly, whether that be forgoing far-flung destinations and finding joy in places closer to us, investing in low-carbon transport, compensating for those unavoidable journeys using carbon offset, or embracing ground-based travel whenever we can. If you take this path and adapt your way of life, perhaps you will even discover a new-found love of the pleasures of slow travel and an appreciation on your own doorstep of what you have been missing.

Manifesto for change

- Prioritize sustainable mobility. This can be done by speaking with your local council about improving infrastructure for pedestrians and cyclists and creating public transport options and more nature and recreational space.
- Switch your fuel. Change your family or company's transport fuel to biodiesel in the short term and aim to move towards electric in the medium term.
- Join the Playing Out campaign with neighbours and friends. This will allow you to organize street play sessions where you close roads to cars on certain days and radically change and enhance your street community.
- Encourage your school/business/company to set emissions reduction targets. First aim for carbon neutrality, then continually increase your targets to become carbon negative year on year.
- Switch to a more plant-based diet:
    - a standard Western meat-based diet produces 7.2 kg CO_2e/day
    - a vegetarian diet produces 3.8 kg CO_2e/day
    - a vegan diet produces 2.9 kg CO_2e/day.
- Stop flying altogether or reduce the number of flights you take. Consider whether flying to your destination is absolutely crucial. If not, look into alternatives and slow travel options.

- If you must fly, offset your emissions. For this, I'd recommend Native Energy, Sustainable Travel International, Terrapass, Clear, myclimate, and 3Degrees.
- Divest your pension from fossil fuels. Change pension funds, or even better, lobby your existing fund to divest from all fossil fuel investments.

Things our governments must do:

- Implement intercity bus networks.
- Tax aviation fuel based on carbon emissions per mile.
- Stop subsidizing fossil fuels for power generation and transport. Instead, exempt clean energy and fuels such as hydro, solar and wind-based power from tax.
- Build new electric high-speed train networks across the world.
- Introduce mandatory food labelling. Make sure that this includes the country of origin, the carbon footprint and the farming method.
- Plan new settlements and housing around accessibility to public transport. This will prevent housing being squeezed into roads that allows cars to take precedence.
- Reopen old railway lines. These were closed in the twentieth century in the belief that train use would decrease.
- Provide bike (including cargo bikes), scooter and skating hire points all around the city.

The list is endless. Transport is one of the worst-polluting sectors, and yet one of the easiest for us to solve. So imagine, now, a future where we see the decline of the car. It is not just a planet without pollution, it is increased social space, civility and adventure. This is the age of reclaiming our cities, our bodies and our health. It is the age of the anthropoid, not the automobile.

Chapter 3

Hungry for Change

Agriculture, deforestation and plant-based diets

Laha lives with his family on the island of Madagascar, off the coast of east Africa. In his fourteen years, he has seen the rains become scarcer, until one year they just stopped arriving. The wind whipped at the land, taking and taking but never giving back. Soon the land hardened and cracked until they were farming desiccated dust. It became a landscape coated with a membrane of bones, sweeping and sunburnt.

Laha's mother turned from agriculture to selling timber, and Laha and his siblings began to scavenge for red cactus, wild leaves and locusts. Laha took up the role of collecting water from the trunks of baobab trees and looking after the zebu cattle. Whatever he did, however much of his childhood he sacrificed to labour, however much care he took with the cattle or the water he wrangled from the baobab trunks, it wasn't enough. Two of his siblings developed severe acute malnutrition, and all of them are feeling the effects. In fact, southern Madagascar is experiencing its worst drought in four decades. More than 1.14 million people are facing food insecurity as a result of the drought being exacerbated by human-induced climate change.

The thin veil that stands between life and death is the question of whether you will be able to nourish yourself. But food relies upon water, and the changing climate means less rainfall, less water, and so less food. This is a reminder that climate change and environmental destruction aren't stand-alone issues to be confronted and tackled, they are powerful warnings written in the language of droughts and floods, storms and fires, hunger and exhaustion.

I think one of the most jolting moments you can experience is when you suddenly realize how marginal, fragile and finite the world is. It is largely a closed system, which means that if someone takes too much in one place, others will be left lacking elsewhere. Therefore, one of the key ways to address the scarcity in Madagascar is to realize that it is caused by overproduction elsewhere. It is the intensive and largely unproductive agriculture of other countries that is contributing to global warming, and consequently to the drought that is stopping Laha and his family from farming.

So let's travel to a different farm for a moment: my uncle's. It is in Worcestershire in the UK, where more than 90 per cent of the total income of sheep farms comes from government subsidies, and yet the government seems stumped as to why our fields bitten to the quick are no longer teeming with wildlife. I spent much of my childhood at my uncle's farm. Some of my fondest memories were formed squatting on mounds of sweet steaming hay under the cold stars as I watched a ewe push and strain until a perfectly formed lamb slid out into the yellow light of the lambing shed. I always felt an incomparable calm sitting alone in the shed on duty whilst the family finished their meal indoors. I was alert, watching for the slightest tightening of a sinew, a quickening of breath or pawing of hooves. It was in that steamy shed that I learnt about birth, and also about the thin line between life and death. Nothing felt quite as shocking as when the lamb I had watched slither into the world so perfectly took a few gurgling breaths and then died. What shocked me even more was how quickly you had to harden yourself to that reality.

Although I soon accepted that not every lamb would make it, even at a young age there was a clear distinction for me between death by accident, and death at the hands of humans. I would help the quivering lips feed on bottles of warm milk and I'd stay for long enough to watch the lambs grow and bounce around the fields in bounds of joy, but I never stayed to watch them go to the slaughterhouse. Although when I returned and the sheds were empty, I knew exactly what had happened. However, I never doubted the ability of my uncle to care for the animals. I admired the long, hard nights and the physicality he dedicated to his work. I admired the way he knew how to bring breath to a lamb on the brink of death and how he would plunge his arm into a sheep and bring it back out with a new life on the end of it. But even at that tender age, I was unusually sentimental, and, like most children, persistently philosophical. I began to wonder why we could love and care for animals and at the same time butcher and kill them. I wasn't vegan, or even considering it, I was just curious at that point.

A few years later, the question still sat burning in my brain. At this time I was obsessed with anthropology, and saw the great ape motivation behind every human action. A man weightlifting at the gym was trying to assert his dominance. If a woman was putting on make-up, she was trying to attract a mate. A baby gripping onto my thumb was doing so in case I began to swing through the trees and it had to hold on. I would recommend you try looking at the world through this lens for a few days. Analyse people's behaviours as though you're Jane Goodall studying the chimps in Gombe. We are so afraid of the primitive, so intent on being civilized, that we often forget that really we are still very much just cave people in suits. We regard ourselves as animals only when it suits us, and as separate when it comforts us, like when we're eating other animals.

During this period, I became disillusioned by the way I saw people consuming meat so readily and frequently. Bacon for breakfast, chicken for lunch and salmon for dinner. It was the consumption without thought that really bothered me. I wasn't opposed to the idea of meat-eating per se; humans have eaten other animals for a long time. Scientists even say that our consumption of meat could have been one of the catalysts for our evolution. When we spent our days running around the savannah in pursuit of our food, eating other animals was the easiest way for us to access a package of calories containing all the fats, proteins and vitamins we needed for brain growth and maintenance. However, in a society where we now have a much more sedentary lifestyle, and our food comes in cans, tins and cellophane rather than from hours of pursuit, we simply don't need to eat in the same way that we used to. Although some say that to be human is to eat meat, to be human is also to have the ability to reason and show morality. Reason tells us that an indigenous man who spends his life learning to hunt and keeping the ecosystem in balance is not the same as the man who drives to McDonald's to eat the butchered body parts of a factory-raised cow whilst never even setting foot outside his gas-guzzling car.

Eating meat has become conventional, meaning it's hard to see the immorality in it unless it slaps you right in the face. In December 2019, I went to a rabbit factory farm in Nantes, France, for a documentary. We all perhaps suspect the truth of where our food comes from, yet we keep it pushed down in the dark places of our mind, disavowed and dusty. Let me describe for you what the reality is very often like.

We walked into a heavy darkness through thick, rusted doors. We could not see but we could feel the presence of thousands of creatures packed tightly into the shed. It was a sensory tsunami of faeces, acrid ammonia and the snuffling of thousands of rabbits in their wired cages. When the bright floodlights spluttered and flickered to life, they illuminated, with great clarity, a broken agricultural system that spoils more than it serves.

We walked down the aisles of cages packed with baby rabbits, mothers, large rabbits, fat rabbits and sick rabbits. All of them white, with startling red eyes and crusty brown ears. The farmer flipped open some lids, plucked out dead babies and tossed them into a pile. We watched the artificial insemination too, but I'll spare you that. It was mechanical and devastatingly brutal, clearly providing for human greed, not human need.

Needless to say, jamming drugged, overstressed animals together is not only detrimental to the animals imprisoned, but also central to the ill health of the planet and people. The farmer became tearful many times during our conversation. Several of his friends had committed suicide, feeling trapped in a system where they must feed more and more people when they barely received enough money to feed themselves. He was adamant that he and his fellow farmers would transition to a more ethical and sustainable type of agriculture if they had the support from the government.

So, many farmers are not faring well, but what about the rest of us? Laha, his family and millions of other people are critically malnourished, whilst billions are morbidly overweight. The food system is responsible for a third of all greenhouse gas emissions. Whichever angle you come at it from, there are deepening cracks in the architecture of this industry that hint at impending collapse. We're fundamentally hungry creatures. Many people base their lives around food, and it is accepted knowledge that everyone's favourite hour at school is lunchtime. Yet the way we feed ourselves at the moment means we're feeding bigger beasts than our stomachs; we're fuelling the climate crisis, malnutrition, obesity, disease and mass extinction.

In September 2019, I met Paul François, who was unlike anyone else I'd ever come across. He had the brawny figure, thick fingers and calloused hands of a man who had been hardened by years of working the land, but his eyes were soft and gentle, deep wells of emotion. He had been through hardships and horror in his life fighting against one of the biggest agrochemical and agricultural corporations in the world, Monsanto.

Paul was like any other farmer, cultivating his crops, trying to create the biggest profit and harvest as much as possible. In 2004, whilst spraying his fields with pesticides, he inhaled some of the chemical product Lasso. He became so sick, suffering splitting migraines, memory loss and stuttering, that he could not cope with the physicality of farming any more. Medical tests found the hazardous chemical chlorobenzene in his body, although the label on the Lasso pesticide stated otherwise. Paul gathered his strength and, like a real-life David, threw himself into a fight against the Goliath of the multimillion-dollar corporation. I had so many questions after hearing his story. How had Monsanto been allowed to betray people and threaten so many lives? How had Paul found the fortitude to continue when his wife died, potentially from the very toxic poison he was fighting against?

Paul ended up winning the battle against Monsanto. It was found to be their fault that he had spiralled into the depths of illness, because he had inhaled a toxic substance they had failed to warn people about. But ironically, you may have been led to believe that pesticides are essential for feeding a fast-growing global population . . .

Two months after I met Paul François, a landmark study broke in the UK. The report was terrifying, but went largely unreported by the media, even though the headline could hardly have been more horrifying: 'Insect apocalypse poses risk to all life on earth, conservationists warn'. It claimed that 400,000 insect species faced extinction. I read a little further and the undeniable truth brought Paul's words flooding back to me: 'Twenty-three bee and wasp species have become extinct in the last century, while the number of pesticide applications has approximately doubled in the last twenty-five years.' Around the same time, I had been driving with my dad on a mild, clear evening. The headlights pooled on the road ahead, illuminating the tarmac and the road markings but little else. My dad was shocked at something that seemed so normal to me. He told me how they used to have to stop on night drives to clean the windscreen because there was such an abundance of insects. Now our windscreens are clean, but a third of insect species are endangered.

All these fragments of information assimilated into one glaring truth: our desire for intense productivity is killing life on earth, making us more unproductive than ever. We are rendering our once-wild countryside into a barren desert. An ecological wasteland. Pollinators have little to pollinate among the vast monocultures of crops dripping with toxins.

For me, the biggest question is: why are we using so many pesticides? Yes, we have a planet straining at the seams with almost eight billion humans, who must all eat to survive, but according to the UN Food and Agriculture Organization, we are able to feed nine billion people today. So what is the real reason for our reliance on these chemicals? It's a question with many answers, but also a question we must try to get to the bottom of if we're going to survive beyond the next few decades. Pesticides are used to increase productivity. So how could we move towards a future where we are just as productive without them?

There are lots of possible strategies. Farmers could try to boost crop yields through new farming techniques or through improved crop genetics. The most effective solution would be if we actually farmed for people, rather than growing food to feed other animals, mainly cows. Livestock and humans now make up 96 per cent of all mammals on earth, and feeding all of those stomachs is a big job, especially since the vast majority of farmland, meat and dairy, accounts for just 18 per cent of all food calories, despite taking up so much of our planet. The charity Action Against Hunger says that for every 100 calories of grain, a cow will produce only 40 calories of milk, chickens will produce only 22 calories of eggs, and we get only 12 calories, 10 calories and 3 calories of meat from chickens, pigs and cattle respectively. Perhaps you're starting to get a glimpse of the inevitable destination we're heading towards with this train of logic.

In October 2019, scientists released a report saying that we need a 'global shift to a "flexitarian" diet to keep climate change even under 2°C, let alone 1.5°C. This flexitarian diet means the average world citizen needs to eat 75 per cent less beef, 90 per cent less pork and half the number of eggs, while tripling consumption of beans and pulses and quadrupling nuts and seeds.' So in essence, if eating meat is a spectrum, we all need to move towards the side of veganism.

Veganism. The word is to the brain what Marmite is to the tongue. It either conjures up joy and hope, or revulsion and denial. How has a philosophy rooted in non-aggression stirred up some of the most virulent fights on social media? For many, the firm opposition to veganism is not really about taste or preference; instead the anger originates when people feel their individual freedom is in danger of being compromised. As our diets are highly personal and subjective, being told what to eat feels for some people like the ultimate challenge to their freedom. Those who hate veganism feel not just that their steak is at stake, but their identity.

Although many approve of the vegan diet, they feel uncomfortable with the wholesale ideology behind it. Some may say 'Okay, I'll eat vegetables and reduce red meat consumption for my health', but the moment it becomes about animal welfare, those same people will post a picture of their fry-up on social media just to incite anger. However, these people are either becoming few and far between or are quietening their opposition in the realization that we are teetering on the brink of great societal change. Being seventeen means I've yet to experience the after-effects of generational upheaval. However, one of the most remarkable changes I've witnessed so far has been the transformation in attitudes towards veganism.

The concept of benevolence and causing no harm towards other species isn't new. Followers of Buddhism,

Hinduism and Jainism have been practising these traditions for a long time, but it was only relatively recently that the tendrils of these movements began to lick at the Western world.

In November 1944, a British woodworker, Donald Watson, gathered five friends and announced that he didn't want to be called a vegetarian any more, because they ate eggs and dairy, which went against his philosophy. Together the group brainstormed ideas for a new label. 'Dairyban!' someone piped up. 'Vitan,' said another. 'No . . . benevore,' proposed a third. Finally they settled on the term 'vegan', the first and last letters of 'vegetarian'.

It was on that winter's evening that the floodgates opened and thousands of years of ideology, philosophy and tradition poured into the West. It began slowly, like the formation of a formidable tsunami in the depths of the ocean, gathering strength and energy. People who were already eco-conscious, slightly hippy, and health-aware began to unite around the idea of veganism, creating an ever-growing group. Groups are one of the most powerful things humans have ever created, but also one of the most dangerous. From families and tribes to schools and bands, agricultural communities, cities and nations, groups govern our lives and make us feel worthy. Often the stronger the ideology behind a group, the tighter it becomes and the more pride it takes in its supposed superiority. That's why, at the beginning, with a shiny new ideology to wear on their sleeve, many members of this new group made veganism feel like an exclusive club. However, documentary films like *Cowspiracy* and *What the Health*, companies like Beyond Meat and initiatives like Veganuary mean those dusty exclusive associations are being shaken off, and modern culture has adopted veganism as a proud label of belonging. Among my age group, to be vegan is not to be a grass-eating caveman; it is a badge of honour.

My relationship with veganism is like that of many other teenagers: a struggle against resistance from parents and family. I've always been small. I look years younger than my age group at school; at seventeen, I can pass for a thirteen-year-old. My parents were hesitant about veganism even before I started. However, knowing that I am undyingly stubborn and would persist secretly regardless, they watched with a mixture of horror and amusement as I dabbled in this new lifestyle. From the ages of twelve to sixteen, I leapt from being vegan to pescatarian, to vegetarian, and then back to vegan again, mainly to conform with my family's wishes. However, I was always drawn back to veganism by the undeniable truth that the Western expectation of having animal products readily available to fulfil our every whim and want is one of the largest drivers of what is wrong with this world.

As food is bound up with personal identity, and is thus inextricable from politics, veganism has unfortunately become a political statement, which immediately puts you at the mercy of those on the opposing side. This means that becoming an outspoken vegan is an undoubtedly courageous move, particularly when you're only six years old . . .

Genesis Butler defies the negative vegan stereotype that has been implanted in our brains. Even today, many people still envision vegans as sallow and sickly-looking, with permanent frowns on their pale faces, scowling at the horror of the world. Add hemp, sandals and a free-loving commune in there too, and that's how much of the population perceive us.

Genesis is thirteen, from California. She is bubbly and humorous, cracking jokes and spilling anecdotes and stories constantly. I have to be honest, I had slipped into the very mindset I spend so much time trying to fight against. I was curious how, when she was still just a toddler, Genesis could have had such a wide awareness of the impact of her own actions and consciously reject something that, if you are brought up being taught that meat is good for you, seems so essential and normal. My mind began to tiptoe tentatively around 'indoctrination' and 'parental persuasion', but when I heard her story, I realized that to an extent, everything we do is a result of culture, our parents and some form of indoctrination.

If children were brought up in an alternative vegan world and then sent to earth, they would be horrified at the fact that we eat animals. Very early on, we internalize the ethics of our culture, but there's a moral clarity in childhood that, even at only age seventeen, I think I've lost in a way. When we are older, we begin to rely on laws and rules to govern our behaviour. When we're young and we feel the car is going too fast, we shout 'Slow down!' When we're older and we feel it's going fast, we will notice we're under the speed limit still and perhaps accelerate a little more. The unflinching distinctions between good and evil that young people make are something we often forget as we get older. The way we used to say to a classmate 'You're being mean' or 'You're not being nice' when we didn't like their behaviour stems from the same moral clarity that made three-year-old Genesis feel unsettled when her mum put a plate of chicken nuggets in front of her.

Her mum, Genelle Palacio, paints a picture of a young child in love with animal products. Genesis would eat only chicken nuggets and nothing else until age three. The age is relevant because it is startlingly young, but also because three to four years old is when children begin to grasp the concept of irreversibility, of something being gone and never coming back. I remember my own first encounter with death. It feels strangely selfish that I even recall this story, because of its inconsequentiality in comparison to the experience of others, but here it is anyway.

I was about four, obsessed with insects and mud pies and climbing trees. I got to the point where the divide between indoors and outdoors was too much to bear. I didn't want to leave the wonder and enchantment of the garden and come into the sterility of the house, so I began to bring insects in and hide them in Tupperware containers. One summer day when my family were all occupied, I snuck in three ladybirds in my small, hot fist and tucked them away in a plastic box in full sunlight. When I came back a few hours later to check on them, I was shocked to find not the vibrant crawling creatures I'd left, but instead three crispy, withered fragments with their wiry legs stuck up into the air. I studied them intently for a while, then ran downstairs wailing, with the brittle beetles rattling in my palm. I remember I was so inconsolable that our neighbour came round to distract me from what I believed was the worst thing I had ever done. How strange that I still remember that immense guilt and the feeling of those hot, regretful tears.

It's clear just how overwhelmed and confused young children can feel by death, whether it's of a ladybird, a cow, a pet or a person. Maybe we never truly come to understand or accept mortality, but Genesis's growing grasp on the idea of death coincided with her realization that some creature had died and was now chopped up in pieces on her plate.

A couple of years later, she was watching her mum breastfeed her sister and asked her why *she* didn't have to breastfeed to get milk, and where the milk she drank came from. Her mum remembers thinking 'If I tell her the truth, she's going vegan *today* . . . Should I do it?' She told her that the milk came from mother cows, and just like that, Genesis decided that it wouldn't be fair if someone took *her* mum's milk, so she wasn't going to drink milk from another creature. She had, at age six, made a decision that was to define the rest of her life.

As many people do when they feel something is undeniably wrong, she began to speak up about it. She says she rarely has to convince people to care about animals, but rather to align their existing care for animals with their actions towards them. She is living her beliefs in a way many people don't. On her first day of school, she wore a T-shirt that said 'Love animals, don't eat them'. When she got home, her mum asked her how her day had been.

Genesis said the other children had thrown hot dogs at her and rubbed salami on her face. 'Were you okay? Did you cry?' her mum asked. 'Yes, I cried,' Genesis replied. 'Not because they were making fun of me, but because they didn't understand they were hurting animals.' She wanted more than anything for them to understand that her lifestyle was not a sacrifice, but actually a development.

I agree with her. Whether it is climate action, wildlife protection or veganism, we have for some reason started to tell ourselves that solving these problems requires sacrifice. But is losing millions of species not itself a sacrifice? What about the erosion of our life support systems, such as rainforests and coral reefs? Are dry, arid cities, withered cropland and conflict over oil not sacrifices? What about the displacement of hundreds of millions of coastal people? But no, we see having to own a more efficient car as a sacrifice! Sacrificing is exactly what we're doing right now on a global scale; animals, peace, our health and much more are being disregarded just to retain our beloved wastefulness.

Social media trolls have tried to stop Genesis. People in the street have tried to stop her. A circus manager has tried to stop her. Here's how a typical conversation between Genesis and someone discrediting the concept of veganism goes . . .

Genesis: Is it morally acceptable to kill humans for food?

Meat-eater: No! Obviously not.

G: If it is morally acceptable to kill pigs for food and not humans, there must be a difference. What is that difference?

M: Clearly I am a person and a pig is a pig, that's enough difference.

G: Okay, so in your mind the main reason is genetic make-up, right?

M: Yes, I guess so . . .

G: Okay, so what if aliens came down to earth and decided it was acceptable to eat humans because we had a different genetic make-up? Would you agree with them?

M: No, but we're intelligent and sentient, so we could tell them not to eat us.

G: Okay, but what about the fact that the high IQ of pigs actually makes them smarter than large demographics of humans?

M: Okay, okay, I understand, but we have been eating meat for a long time.

G: Humans also kept slaves for a long time, and in many parts of the world we are at a point where we can fulfil our biological needs without animal products. Tradition is a poor excuse for continuation.

M: BUT I LIKE THE TASTE!

Essentially that's what it comes down to. The meat industry is driven not by necessity, but by temporary pleasure, which could arguably be matched by plant-based alternatives. Some may well find these arguments, and vegans in general, too evangelical or preachy. But the thing about vegans that really unsettles people is not the bombastic lectures on why we should give up animal products; it is that they have a good point.

In every video I watch of Genesis talking to an audience,

she has to stand on a box or a stool in order to look over the podium at the sea of people in front of her. It's moving to know that this young girl who still refers to her one-legged teddy as a person is the youngest to give a TED talk, has challenged Pope Francis to adopt a vegan lifestyle during Lent – and successfully received a response – and has set up her own foundation to care for farm animals. She has taken up a noble fight at a young age and has a thorough grasp of what she's talking about, with the statistical realities of soy, cattle and deforestation slipping off her tongue with ease.

We've all heard these narratives being pushed before. Hearing from a vegan in a major city is one thing, but what if we follow that thread and meet one of the young people who live in a place where the cows have yet to become steaks.

The Amazon: an explosion of life and a cornucopia of colours. A heaving organism thrumming with the calls of species we've yet to know and possibly never will. A hub of the most elemental substances: water, soil, foliage . . . and fire.

Right now, as I sit writing these words, a large swathe of our planet is quite literally burning, and almost no one is speaking about it. The fire season in the Amazon has just begun, and is worse than ever. The fires burn hard and spread fast, churning out acrid smoke that turns the daytime skies 2,500 kilometres away in São Paulo as black and dark as night. The lungs of our earth aren't short of air; they are being choked, and humanity is the hand around the planet's throat, taking life and bringing death.

The fires are caused by us, whether directly or indirectly, because nearly 50 per cent of cattle bred for consumption are raised in fields that used to be rainforest, which has been burnt to the ground to make way for the cows. Many people angrily tweet and post on Facebook when the uncontrollable blazes ravage the rainforest, then get in their car and go and buy a burger or a steak. We're confused and enraged when we see a torrent of apocalyptic images of caustic smoke smothering the lush green of the Amazon and cities as far as 1,700 miles away . . .

Artemisa Xakriabá doesn't live 1,700 miles away. She lives in the beating heart of the crisis. I first heard about Artemisa in September 2019, when another spate of fires was sweeping through the Amazon. I was at a climate strike in London and she was in New York. Four million people around the world were protesting that day. There's a very powerful feeling of unity at a protest, and I'm convinced this feeling is a strong motivator for mobilization. The reticence of daily life dissolves and it's like you all become one throbbing, chanting organism. In a way, the word 'protest' is wilfully wrong. Events like the climate march aren't just about hating the system, the government, or what is happening in the world; they are also largely about love. Solidarity could be said to be the basic, most pressing social need of humans. Little feels more unifying than the experience of collective action and building something new together as the same chants flow from the mouths of millions of people who've never even met before.

I sat on the train on the way home absorbed by the surreal silent hum that follows the breathless enthusiasm of a protest. As usual, my feet were blistered from the chokingly tight Converse I insisted on wearing to every march. My sign was battered and shredded from being thrust into the air and my stomach was growling from hunger, yet I felt strangely happy scrolling through the #ClimateStrike hashtag, looking at the millions of people around the world who had spent their day just as I had. Another young person giving a speech at a climate strike popped up on the screen, and although I had heard many that day already, I pushed my headphones into my ears to listen.

'My name is Artemisa Xakriabá. I am nineteen years old, and I am from the Xakriabá people in Brazil. I am here today representing the more than twenty-five million indigenous and traditional communities from the Global Alliance of Territorial Communities.'

The Xakriabá people have to contend with some of the most disastrous effects of humanity's war on the environment. Many of them live in the Cerrado, a sweeping tropical savannah bordering the Amazon that has thousands of species that cannot be found anywhere else in the world. However, over half of its landscape (grassland scrub and dry forest) has been converted to agriculture, as it produces soy for markets mainly in China and Europe. In the last decade, the region has lost 50 per cent more than the Amazon, which has more legal protection.

Where the Xakriabá see home and an abundance of life, others just see an abundance of resources. I will never understand this, no matter how much I read into psychology, capitalism or consumerism. How can one human regard a swathe of land as something precious and worthy of respect, and someone standing right next to them view it only as capital to exploit and benefit from? Whatever the answer, the latter means that the home of the Xakriabá is now one of the most devastated areas in the world, with only 20 per cent of its original coverage.

'We are fighting for our sacred territory,' Artemisa said in her speech. 'But we are being persecuted, threatened, murdered, only for protecting our own territories. We cannot accept one more drop of indigenous blood spilled.'

Another member of the Xakriabá tribe expanded on Artemisa's words in an interview with the *Guardian*: 'When the forest is burned, the birds and the animals, they are either burned or they go away. And this doesn't affect only the animals, but it also affects us. We rely on them to eat. So with no animals, we have to rely on food from the outside, and this ends up making our children and our women sick, here with the Xakriabá people. I can hear the song of the birds now, but it's also a song of misery, of sadness, because most of them, they are alone. They have lost their partners. The birds, they usually sing as a couple. And many of them are now singing alone. And we, the indigenous, are becoming more alone, because they're taking people from us.'

Although 12,000 people live in Artemisa's home town, at just seventeen years old, she travelled 5,000 miles alone to represent her community at the climate strike. She also went alone to address members of the US Congress and the House of Representatives leader, Nancy Pelosi, in Washington. The reason she travelled alone? Back home, the indigenous communities were busy fighting the devastating fires that had been burning for months. They didn't have the privilege of marching or striking for future generations. They were literally in a battle to save their own lives.

'The Amazon is on fire,' Artemisa continued. 'The Amazon agonizes year after year for the responsibility of the government and its destructive policies that intensify deforestation and drought, not only in the Amazon, but in the other five Brazilian biomes. Climate change is a result of this, and it also helps to make the fires stronger. And beyond the Amazon, there are the forests of Indonesia, Africa, North America, whose suffering has such an impact in my life and in your life.'

Artemisa believes that Jair Bolsonaro, the Brazilian president, is trying to put her tribe 'into extinction'. Whilst students around the world were plotting the large September climate strike, Bolsonaro was working on creating a climate of impunity where regulations are weak, criminals are exempt from punishment, and ecocide is rife. He likes to say that by forcing indigenous peoples into the city, he is making them more human, but Célia Xakriabá, another young woman from the tribe, says that 'we are living in a moment of disputes, disputes of values. Leaders like Bolsonaro are like boils on your skin, and they emerge with all the fury, these boils, like a cancer to these values of life. They emerge with this fury because they appear to have the desire to extinguish all diversity – the diversity of life, the diversity of culture, the diversity of seeds, the diversity of territory.'

In the year of the climate strike alone, deforestation in

the Brazilian Amazon increased by 70 per cent. Bolsonaro has been recorded saying, 'It's a shame that the Brazilian cavalry hasn't been as efficient as the Americans, who exterminated the Indians'; and before he became president, 'If I become president there will not be a millimetre more of indigenous land.'

These are chilling words coming from anyone's mouth, but especially from a man who holds such power. As indigenous people make up less than 5 per cent of the world's population but protect over 80 per cent of its biodiversity, an assault on these tribes is also an assault on the planet.

Like many young activists, Artemisa's fight wasn't always such a perilous one. Her journey didn't begin empowering millions at strikes and addressing the US Congress. Rather, it started with soil-covered seven-year-old hands pushing saplings into the ground near her home. Along with other young members of the Xakriabá tribe, she was helping to reforest fifteen riverside areas in the south-eastern state of Minas Gerais. She was brought up very aware of the natural world, and sees humans as part of nature scientifically as well as philosophically. We are mineral creatures, and there is a geology of the body as well as the landscape. Our teeth are like reefs, our bones like stones as we convert calcium into our skeleton in order to function. It's a different way of seeing the world, and of seeing ourselves.

As Artemisa grew, she realized the larger truth about what was happening to her home and family. In distant, barely imaginable lands so different to the one she inhabited, people were consuming more and more animal products. Imagine the confusion for a young girl who got her food from hunting, fishing and gathering trying to understand why the landscape of her childhood was being ripped apart and having to watch helplessly. All of this destruction happening just so that huge corporations could grow soy for their cattle thousands of miles away.

How much of a stupor are we in to not see the absurdity of this situation? We use more land to feed domestic animals than to actually feed ourselves. Considering how much land and energy it takes to keep one cow alive, we may consider them the crudest and most inefficient creatures in the world. But blaming cows for climate change and deforestation is like blaming a baby for causing population growth. The truth is, however many saplings students plant and however many speeches young people give at marches or to Congress, under Bolsonaro's deadly policies and the current rampant consumption of animal products, illegal deforestation will continue, more indigenous blood will be spilt, wildlife will be left homeless and increasing amounts of CO_2 will be pumped into the atmosphere. Unless, that is, we change our ways.

I wanted to speak to someone who had found a new way of living. I called Iris Fen Gillingham, who lives in the Catskills (a province in the wider Appalachian mountains), during a heatwave in London. The grass was brittle, the roads were melting and the air was almost too heavy to stand up in. There had just been a catastrophic explosion in Beirut and a huge oil spill in the Indian Ocean. The world was reeling from the death of George Floyd and coronavirus cases were once again soaring. It felt like the foundations of whatever strange stability we had been balancing on had been knocked down in one fell swoop. It was refreshing to see Iris's smiling face on the video call.

Iris is one of those people who builds something small, something idealistic that works, and then lets change ripple out into the world and touch the people and places it needs to. She also happens to be one of the most sincere and genuine humans I've met. I say 'human', and not 'activist', because, like me, she feels uncomfortable with that label.

'Why not "activist"?' I asked her.

'Well, what is activism? The way my family lives is a form of activism. As a teenager, media outlets began to call me an activist, although I never saw myself as that. If we are just living according to our values and consciously thinking about our impact on the world, that's simply an ethical and just way of living. If everyone tries to live with that certain amount of respect for our neighbours and resources and the land, then we don't need token activists. In our society we spend so much time labelling people and putting people into boxes. Let's just all aim for this value-based way of living.

'Until I went to college, I had no idea how different our way of life was. For me, making dinner meant going out onto the farm, picking the vegetables, washing and cooking them. It just never occurred to me that there was any other way. Almost everything we eat we grow and store on the farm.'

'So you're self-sufficient?' I asked.

'Well, no, we buy rice . . .'

Iris's great-grandfather had also been a farmer; a horse and dairy farmer, who had had a deep connection to the soil and worked the land in a very different way to what we now define as farming. Iris's dad grew up on a farm bought by his parents and then fell in love with someone who was equally as passionate as he was about bringing healthy home-grown food to the community. Together they continued the work of Iris's grandparents, naming the farm 'Wild Roots' and starting a family there.

When Iris was six years old, the family experienced a severe flood, the type that should only occur once every hundred years. However, humanity has always dealt with wild acts of nature, and flooding is, after all, a natural phenomenon. The family were resilient, the farm was repairable, and they picked themselves up and began to clear the damage. A little while later, though, the heavy rains returned in full force. The rivers began to boil and bubble and the farm faced a second hit as the water once again began to pound the soil and lick at the foundations of their family home. The *second* hundred-year flood in the space of just a few months had hit the Catskills. Something was wrong. This shouldn't be happening. Even at the age of six, Iris understood the gravity of the situation. Weary and exhausted from repeatedly fighting to protect the farm, the family once again began to repair and restore their life. Perhaps in an alternative world the story would have ended there. Iris would have noticed something was wrong but would have been absorbed back into the blissful oblivion of childhood. But that wasn't the case.

The third flood to hit was the worst. After picking up your livelihood, land and family from two natural disasters, imagine watching the rain begin to cascade in sheets and pound on the windows again. Imagine the feeling of fear that would strike as the rivers began to swell, rush and roil towards you. Imagine hearing about people you knew plucked from their homes by the muddy waters like a doll at the careless whim of a violent toddler. The third was classified as a 500-year flood (which has a 0.2 per cent chance of occurring in any given year), and it completely wrecked the farm, sweeping away topsoil, greenhouses and crops and destroying the business. The family went from feeding the town to barely being able to feed themselves. Innocent young Iris, with her mop of chestnut ringlets and big dark eyes, had just had her first serious collision with the climate crisis.

Things settled down for a while, but by then Iris was already entwined with something much larger.

Iris's grandmother had wanted her children to grow up wild, with space to roam. Whilst looking for a location for the farm, she had stumbled across a hundred acres of land with a deep well. She decided to buy the land as soon as she tasted the water. It was so soft, sweet and pure that she wanted her children, grandchildren and their descendants to be able to taste it too. Several decades later, eight-year-old Iris watched with horror as fossil fuel companies started fracking in the region. Fracking causes natural gas to migrate into the water supply; as shown in a famous scene in the film *GasLand*, local people could bring a lit match to the end of a hosepipe or tap and blooms of fire and smoke would erupt. Her dad actively started fighting against the companies' encroachment. With vivid images of those terrible floods etched into her memory, and seeing the sweet, soft, pure water being polluted, Iris began to realize the hard truth that to protect the water, the farm and her future, she must do something. 'It's a powerful metaphor,' she tells me. 'What kind of water do we want our descendants to drink?'

She joined two climate organizations, Earth Guardians and This Is Zero Hour, and also helped to arrange the largest youth-driven lobby day and climate march in Washington, DC. However, she became tired of railing and fighting against the system. She wasn't simply burnt out; she had a desperate hunger for hope rather than despair. She said, 'If I wasn't hopeful for the future, I don't know what I'd be fighting for.'

Some of us are cut out for conventional activism, in the form of marches and protests, but this doesn't work for everyone. We must redefine what it means to be an activist so that it is not simply a sign-toting protester, but instead can be inclusive of anyone – be it a farmer, lawyer, singer or artist. Becoming an activist is akin to being a small fish in a river inhabited by big sharks. You can spend your life trying to fight the sharks, but sometimes you have to realize that the best thing to do is gather all the other fish, swim upstream and find a better world. If the sharks can't nourish themselves by eating you, they'll soon perish. So that's what Iris and her family are doing. In an attempt to create a better world, they are working on a new model, and not only imagining the future, but actually creating it.

Iris explained, 'The pathway to a "better world" shouldn't be a plan, it should be a conversation, because after all there is never going to be a perfect world. The process of change we're going through at the moment is just as important and can be just as beautiful as the product we're striving for.'

It's not agriculture that's the problem, it's the way we do it, and Wild Roots is a model of how agriculture could be, with the revitalization of less intensive, but often more productive and definitely more sustainable methods. It sometimes feels like huge, intensive industrial farms are the only way to feed humanity, but small farms of around twenty-five acres or less already produce over 70 per cent of the world's food.

In September 2019, 4,000 miles from Iris's home, I visited another small, family-run farm while filming a documentary. Ferme du Bec Hellouin is nestled in the verdant French countryside of Normandy, and if you want a glimpse into the future, I suggest you go and visit. The moment we turned into the driveway, everyone was captivated by the bucolic magic of the place. No fossil fuels or pesticides in sight, but plenty of dogs!

Coming from London to the farm created a stark contrast. Sometimes I love London, but the feelings of claustrophobia caused by its streets and surfaces of glass, brick, concrete and tarmac were replaced here by a beautiful, explosive disorder of colour, taste, meadows, fields and ponds. Whilst the crew were setting up the cameras, we walked through the greenhouse eating some of the fresh produce (with permission!). Everything was perfectly imperfect. A profusion of strange textures and exotic shapes, yet nothing was thrown away simply because it didn't appeal to the eye of a consumer.

The farm was founded by Charles and Perrine Hervé-Gruyer, with the help of their four young children. The family follow the principles of permaculture, 'a design system for humans inspired by nature'. They aim to re-create the diversity and interdependence you would naturally see in an ecosystem, rather than the desert-like monocultures that dominate modern agriculture today. The farm is only a few acres, but it is the source of a number of scientific studies, which show that it's possible to make a living from such a tiny plot of land by growing food using permaculture techniques.

When you think about it, this seems like a glaringly obvious solution. Nature has persisted in creating an abundance of life for millions of years without oil, tilling, mechanization or chemicals. Nature's circular model produces no waste, and every element is beneficial to the others.

When you ask children what the future looks like, many suggest flying cars, robots and virtual reality. But the children of Bec Hellouin know they are living in a pocket of the future. Nearly a thousand different varieties of crops are grown on the farm, which is the exact opposite of industrial agriculture. Where modern agriculture sees soil as a substrate to pour fertilizers into, the Hervé-Gruyer family see it as the foundation of all life. The soil is never tilled and never bare. They plant their crops and cover them with a mulch that decomposes year after year, enriching the soil. They put the crops very close together, and by planting a dense variety in fertile soil, they can produce up to thirty times more than industrial agriculture could in the same amount of space. This is because you never find monocultures in nature (even areas with one dominant plant species still have other species growing under and around it).

The solutions we are trying to find don't necessarily need to be created by us; many of them exist already and we only have to look at nature to find them. Charles and Perrine have achieved incredible productivity not by some superhuman feat, but instead by listening to our greatest teacher: the natural world.

Those few days we spent at the farm are like a hazy dream. We were residing at the heart of an indescribable privilege, one that makes me so excited for the future. We're sold tales by the wellness industry that contentment is all seaweed wraps and spin classes, but wellness shouldn't be about chasing the next fad diet, detoxing, or constantly trying to reinvent ourselves. It was really moving to see not just the abundance of crops and pollinators, but how this family have found a new way of life and a personal philosophy that has brought them together and created a little slice of hope. They haven't changed the world, but, like Iris's family, they have changed *their* world. Although they seem to be dreaming up a new way of living, they are pragmatic in their quest. Theirs is a dream that must become a reality, and to see them making it happen was an honour. It goes beyond sustainability and is a non-confrontational, gentle rejection of the structures that currently exist and govern our lives. They have refused the unbridled consumption of modern life that is supposedly our path to happiness, and instead presented a new path where we share enthusiastically, live respectfully and find meaning that is compatible with other life forms. Rather than being overwhelmed by the complexity of the problem, they have fallen in love with the creativity of the solution. And that's what we all must do.

For the environmental crisis, some say veganism is the answer, some say recycling, or the circular economy, or changing our mode of transport. But there is no one route to achieving the world we want. Maybe plurality and a diversity of creative solutions is the answer. Indigenous communities aren't vegan, which works for them. Iris isn't vegan, because she's not against meat-eating. She's against the industry that turns animals into 'just meat' and creates such a large physical and social distance between the slaughterhouse and the dinner table that the modern consumer is completely oblivious to the reality of the lives behind their food. In London, veganism is easy and ethical, which works for some of us. The world we need is going to be a messy, diverse patchwork of solutions that differ in every community and culture. We're on a clumsy journey with not just one route, but many intersecting ones. Iris's story highlighted just some of them for us. The possibilities are endless.

Manifesto for change

- Urban farming. The areas where most people live and eat have been rendered food deserts. Isn't it absurd that 60 per cent of London is classified as public open space, but the majority of this has been turned into lawns and ecological wastelands that sometimes require more equipment, agricultural toxins and fuel than industrial farming to maintain? Urban agriculture in London is a tiny enterprise that, if we wanted, could transform the way we feed ourselves. Here are a few suggestions of ways to incorporate urban farming into your everyday life:
    - visit a community garden
    - gather a group in your local community and turn a vacant lot, verge or lawn into a community garden
    - support restaurants near you that are sourcing ingredients locally
    - learn how to grow your own produce, whether that means renting a vacant lot or using plant pots on your balcony.

- Adopt a more plant-based diet. If you live somewhere veganism is a viable option, you should try gradually reducing your meat and dairy intake and move towards that end of the spectrum. Some might prefer to adopt an entirely vegan lifestyle.

- Reduce the amount of food you use and throw away. Plan what meals you're going to eat in your household and do meal prep in advance to reduce waste. Shockingly, the UN has calculated that if food waste were a country, it'd be the third largest global greenhouse gas emitter after China and the US.
- Buy foods that meet a credible certification standard. We all have to become more aware of where and how our food is produced. It's tricky at first, but once you know the brands and products that are certified as sustainably produced, you will find it much easier. Logos to look out for include Fair Trade (protecting farmers and workers in developing countries), Freedom Food (animal welfare), MSC and ASC (seafood), and RSPO (palm oil).
- Demand that governments:
    - encourage farming methods that work in harmony with nature, such as permaculture, agroecology, circular agriculture and organic farming
    - support smallholders by creating policies that protect family farmers and incentivize doing business with them

- financially incentivize farming to restore soils, practise biodiversity and minimize water and air pollution
- tax non-organic pesticides and synthetic fertilizers, which are incredibly harmful to bees and other wildlife.

Chapter 4

Water and Our Worlds

Melting, warming and submersion

Climate science has struggled with a messaging problem. The facts are undeniable, but more often than not, they are ignored. Between the statistics and the lengthy reports, however, sometimes something very strange happens and people actually sit up and listen. Occasionally the climate scientists use a poster child they know will tug on all our heart strings: the polar bear.

A polar bear desperately clinging to some shrinking ice, a polar bear floating adrift, an emaciated polar bear picking its way across a diminishing landscape . . . You name it, the list goes on. Today, it seems almost impossible not to associate these bears with climate change. Yet it's deceiving to tout them as the sole victims of the climate crisis and to pour our mourning into just one species, overlooking the rest of the affected species around the world – including, of course, humans.

'It's cold in here, very freezing,' are the first words I hear down the crackly phone line from Tashi Lama. He's calling me from the village hospital, because it's the only place with a Wi-Fi connection. Tashi is part of the Sherpa community, an ethnic group indigenous to the Himalayan region, whose northern borders are dwarfed by Mount Everest. He lives with his mum and dad in Ghunsa village, 3,500 metres above sea level. It's a three-day walk from the nearest town through the rugged yet graceful arcs of mountain giants, savaged and finessed by wind, water and snow. I was put in touch with him through the Glacier Trust, who are helping indigenous communities adapt to the ever-changing environment of the Himalayan mountains.

The Himalayas provide food for 1.3 billion people living in downstream river basins, and Sherpa communities like Tashi's are the most vulnerable because every aspect of their livelihood, culture and food sources relies on the environment surrounding them.

'In my area we can't grow rice, millet and other crops like that, so we use our potatoes to barter and exchange them for other crops from the lowland. However, last September, the long rainy season caused by the changing monsoon pattern was disastrous for the summer crop of potatoes.'

'So what did you do?' I ask. 'How did you get through it?'

'Some of my village went on a four-day trek to Yangma village, braving ice and stiff winds. We took as many yaks, mules, donkeys as we could and they gave us potatoes, but it wasn't enough to feed everyone . . . we can't continue like this.' He hesitates. 'As well as potato farming, we have another cultural tradition: yak herding. However, the change in temperature means the alpine meadows we use for grazing are being lost as the treeline gets higher and higher and shrubs encroach on the land. Snow leopards are also attacking more yaks as the meadows are being squeezed into smaller areas.'

Tashi travelled to Singapore to study. During this time, he was taught about greenhouse gases, climate change and the fact that in 2009, the IPCC announced that the Himalayan glaciers would melt by 2035. He told me that many of the villagers believe the melting glaciers receding before their very eyes are caused by the gods, who are angered by the way foreign tourists trample and desecrate the sacred mountain grounds. Eighty years ago, a famous Sherpa Buddhist predicted that much attention would come to be focused on Mount Everest, and that people would 'suffer hardship as a result of negative deeds generated in her vicinity'. Everest, or Chomolungma, which means 'the Mother Goddess of the World', is believed by the Sherpas to contain the spirit of a deity, and before any climber sets foot on the mountain, whether they're Nepalese or not, they must ask for permission in a ceremony called the puja, where they request 'a safe passage to the summit and protection from harm while they tread the Mother Goddess's hallowed ground'.

Those Sherpas who don't want to be involved in the hard labour of potato farming turn to climbing and tourism. When you hear of a Westerner who has climbed Everest 'solo', they haven't. They've been guided by the steady hand and wisdom of the Sherpas, who know that the stakes are high and that many will die on the mountain. Tashi hasn't climbed Everest himself, but several Sherpas he knows have done so multiple times; they can haul around 45 kilograms of climbing gear and resources up the mountains to help the tourists.

In 2017, one of the Sherpas saw something pink amidst the naked white expanse of Camp 1. After closer inspection, they discovered it was a human hand from a body that had been buried and preserved for many years. As the fast-melting glaciers drip away, the shining, silent, elemental landscape spits out bodies that have been hidden for a long time. It's like some sinister symbolism of what's to come – the death of the glaciers mirroring our own mortality.

'And . . .' I hesitate, unsure of how much worse it could get, 'are you experiencing any other tangible effects of the climate crisis?'

'Yes,' Tashi says. 'The lakes are exploding. It is getting more and more common and there is never any warning about when it will happen.'

After some research, I realized that the exploding lakes Tashi was referring to are what climate scientists describe as 'glacial lake outburst floods'. They affect hundreds of thousands of people all over the world. The lakes in glacial landscapes are dammed by ice. Water flowing over the ice widens gaps in the dam, so that even a tiny disturbance can trigger a massive flash flood that races down the slope, tearing and ripping at villages like a rabid dog.

Exploding lakes, bodies being exposed by melting ice, hungry communities and even hungrier yaks are all symptoms of the climate crisis at high altitude. The problems don't stop here, though. All of this turbulent water then tears down the mountains until it reaches the ocean . . .

One late August day, I awoke so early that the morning was still untouched by birdsong and the glow of day was only just beginning to rub its muzzle on the window panes. I stumbled down the stairs, opened my laptop and clicked on the Zoom call. I waited, and waited, and after a burst of fuzzy interference, Belyndar Rikimani burst onto the screen. She was calling from the South Pacific, which was eleven hours ahead, and because she had no signal or connection of her own, the call was being managed by the PISFCC (Pacific Island Students Fighting Climate Change). Belyndar has one of those joyful, warm smiles that reaches her eyes. She waved at me with an arm encased in a thick cast, and I waved back. No sound yet, just symbols of greeting. Finally in an explosion of noise, I heard her microphone crackle to life, and she told me her story.

Belyndar, a student, had grown up in the Solomon Islands, a nation comprising hundreds of islands in the South Pacific. Her province, Malaita, is the type of place that would fulfil an average European's definition of paradise. It has fine powder beaches with sighing blue seas that gently lap at the roots of the whispering palm trees. Perfection to an outsider perhaps, unless you've read the 2018 IPCC report on 1.5 degree warming and noticed the small section that explains how without drastic action being taken globally to reduce greenhouse gas emissions, Pacific island countries will cease to exist.

Belyndar grew up in the mountainous region of Malaita, so at first, the waves lapping ever closer weren't a worry to her. In fact, she didn't even notice. What she did notice, however, was that the rivers around her house had become mere trickles rather than torrents and it was taking the village women longer and longer to get back from their daily water-collecting trips.

Later, she travelled to the edge of Malaita to volunteer

and help coastal communities who had been battered by the encroaching waves. There's usually a distinct point when the subliminal awareness of environmental destruction suddenly bursts from the edge of our consciousness into clear focus, gripping our whole being with fear and shock. That was the moment it happened for Belyndar. Many of the people of the Solomon Islands are human mermaids. They plunge into the ocean as young children and spend much of their life fishing, rowing and immersing themselves in the watery world of crustaceans, molluscs and coral. When Belyndar turned up to volunteer, she saw the communities treating the ocean like a rabid beast. The parents would tentatively watch their children, and as dusk began to fall, there would be no more splashing and playing. Instead, the children would be ushered away from the water's edge.

She questioned the village women about this unspoken anxiety and found out that the communities were afraid of 'king tides', which swell up suddenly and suck people out to sea. They're a result of the alignment of the moon, sun and earth, and have been happening for as long as tides have existed, except now the rising sea levels and increased frequency of storms were nourishing the beast and creating something monstrous.

When Belyndar returned home, she mulled over what she'd learnt. She began to worry more about what would happen. During her anxious musings, staring off into the horizon, she realized she was looking right into the eye of the crisis: Walande Island, an artificial island made from coral rocks that have been piled up over hundreds of years. It was originally populated by people who wanted to escape the mainland, which was the breeding ground of mosquitoes and malaria.

Walande Island was visible from Belyndar's window, but it was also a window in itself. A window into the future. The sea-level rise around the Solomon Islands is three times faster than the global average (three millimetres per year). The higher local rate is a result of natural climate variability, but gives us a good indication of what will happen if we don't dramatically slash greenhouse gas emissions. In other words, the normal of the Solomon Islands is very quickly going to become the normal for the whole world. It is not only the rising sea levels that are having such a detrimental impact on these islands, but also the energy and sheer power with which the waves crash against the shore. The warmer atmosphere creates more energy for them, until they are like giant hands, becoming ever more nimble and dexterous as they pluck away bits of earth, homes and lives.

Belyndar had never been to Walande Island, but had spent a lot of time watching it from the window. She began to hear word that the inhabitants were in trouble. That people were being pushed off the island. The year I was born, a *Blue Peter* documentary aired about Walande Island; it would become, unknowingly, a sort of time capsule documenting the last few years of a place and its culture. In the documentary, the presenter navigates towards the island in a boat. As it approaches, you see the locals rushing towards the cameramen with bows and arrows, cautious of the unusual equipment being unloaded.

A few moments later, the crew are being welcomed by a pulsating dance of loose limbs, harmonicas and traditional headdresses. At the time, there were 1,200 inhabitants and it looked like they were squeezing life to the bone, grinning widely and cheering their new visitors.

Fifteen years later, the filmmaker who had made the documentary in 2002 returned to the island to film again. You see his boat weaving through wooden remains that lunge out of the ocean like abandoned tombstones in a graveyard; the skeletal remnants of a submerged society. Only one house survives. It belongs to a man called Timothy, who lives there with his grandchildren. The other villagers have all moved away, and when they're asked about Timothy's reasons for staying, they shrug indifferently. 'He's stubborn,' one says. 'He won't listen to us,' says another.

Every morning his four young grandchildren take the canoe to the mainland, where they go to school, and Timothy spends the day adding rocks to the wall around his house, trying to hold off the water for a bit longer. 'If I move to the mainland, I can't see anything through the trees. I won't even see the water. I want to have this spot where I can look around me. Because I'm part of this place,' he says.

When Belyndar learnt about Walande disappearing, she tried to imagine permanently leaving her home. This isn't just like moving house. It is leaving behind a society and everything you've ever known. You are leaving the familiar earthen odour of the island, the familiar angle at which the copper sun sets on the horizon, the familiar ambient splash of water against the fishing boats in the silver twilight. You are leaving a place pockmarked with centuries of ancestral memory and history. You're leaving part of yourself and losing it forever.

For Timothy and Belyndar and the women on the Malatian coast, climate change is not an academic debate at the dinner table; it is literally watching your dinner table crumble into the ocean.

My Zoom call with Belyndar had malfunctioned several times by this point. Somewhere on the other side of the world, a young woman was turning the connection on and off, trying to re-enter our meeting room to continue her story. Somewhere on the other side of the world, islands were being swallowed by the sea as I sat there in the pale morning light of London. It seemed strange that one of us could live out their life however they pleased, not caring that their actions were having a catastrophic impact on the life of someone they'd never meet. It all seemed so . . . Belyndar burst onto the screen and derailed my bleak train of thought.

I wanted to know what you do after an event like that, and I wanted to know if Belyndar was frothing with righteous indignation at just how unfair the whole situation was. It's like when a dog mauls someone: the blame usually lies with the owner. The violence and theft being committed by the encroaching waves is almost entirely the fault of humanity. Species become extinct, islands are submerged, lives and traditions are lost and forgotten about as easily as coins down the back of a sofa. What we need to fight is not the change itself, but our careless actions; science has shown us that the baseline of extinction, sea-level rise and global temperature has been shaken up by humans. Extinction rates are a thousand times what they should be without our interference; we are essentially committing genocide on a taxonomic level.

'Yes, I feel emotional when I see what's happening to the animals and to my people,' Belyndar told me, eyes glistening. 'We're living a day at a time here, and even though we might be content with today, tomorrow is always built on uncertainty. The most painful thing is to see Western culture going along with work as normal, and at the same time, seeing my people suffering so much. When I see how people are suffering from something they have not contributed to, it's a very sad reality.'

Belyndar's journey began a few years ago, when she was studying law. She met another young law student, a man called Solomon, in a café and they began to talk about climate. It turned out that Solomon was assembling a group of law students to take climate change to the International Court of Justice and to educate other young people. They both fell into hearty agreement that the smirks of politicians at the Paris Agreement and the lofty promises of the UN stank of hypocrisy. For Belyndar, it was a relief to speak to someone who saw the cracks in the facade, and as she left, she told him she would attend the group's next meeting. It was to be the beginning of the PISFCC.

What they're fighting for is legal change. Submersion for many at this point is a fact, not a prediction. It is painful to say, but the truth often is. According to the IPCC, the whole world must be carbon neutral by 2050 if we have any chance of limiting warming over 1.5 degrees, preventing runaway climate change, avoiding the displacement of billions and losing many ecosystems all over the world. So now consider this: 71 per cent of industrial greenhouse gas emissions are linked to only 100 perpetrators, including ExxonMobil, Shell, BHP Billiton and Gazprom. If we, as a global community, listened to students like Belyndar and Solomon, who are advocating for a law that supports harmony not harm and puts people and planet over profit, then imagine what could happen, imagine how quickly we could effect change.

This mass damage to and destruction of ecosystems globally is known as ecocide. Remember that word and use it as much as you can. It needs to slide off our tongues and be understood as something catastrophic, just like genocide and war crimes. Genocide, war crimes, crimes against humanity and crimes of aggression are the four existing international crimes against peace in something called the Rome Statute. Mass environmental destruction was always meant to be the fifth crime against peace; it was even included in the drafts. But to the horror of anyone who recognizes that ecocide is a crime against the peace and security of humankind and the natural world, the final international version had no mention of this fifth law at all.

Now we must travel from the Pacific all the way back to the scandalous, flamboyant 1960s in the UK. The swinging sixties were a liminal time, a decade of art, music and fashion. Up in Scotland, the Highland Boundary Fault was a liminal place, a great crack in the land that separated one terrain from another. This is where Polly Higgins came into the world. She too is like the sixties, or the Boundary Fault, because there is a 'before' and an 'after' Polly Higgins, so profound was her impact on the world.

Born in 1968, Polly can safely be considered a child of the Anthropocene because she was fighting injustice and extinction as a child and never stopped. It began sitting at the kitchen table as a young girl. She would hang on the words of her father, a meteorologist, who warned of something impending, a change in the atmosphere and the climate. A concern for the bigger picture was instilled in young Polly just as some parents embed ambition or apathy in their children. But something that was already inherent was Polly's outrage at injustice. At school, she observed a teacher abusing a fellow student repeatedly; one day, when another blow was about to be inflicted on the young pupil, she stood up and punched the teacher. She was expelled, but no one could expel her spirit of justice, which eventually led to her becoming a lawyer.

First she delved into art for six years, living in London and dealing in old paintings. But even though she hated public speaking and was deeply shy, some intangible calling drove her to the law. Her mum told her not to do it, her dad told her not to do it, but she became an employment lawyer and then a barrister, dealing with heavy courtroom cases. A while later, at the end of a three-year case representing a person who had been injured at work, Polly sat looking out of a window at the Court of Appeal, watching trees being cut down outside. The earth is being injured as well, she thought, and yet there is no one there to hold the perpetrators accountable. The voice of the young Polly who had stood up and defended a fellow pupil when she saw injustice began to get louder.

It's an injustice to kill people, so we criminalize it. It's an injustice to steal, so we make it a crime. Likewise, it's an injustice to cause mass damage and destruction to the planet, so we should criminalize that as well . . . The earth is in need of a good lawyer! Many of us have these lofty visions of change, but few of us reflect on what needs to be done and decide that we are the ones who must do it. Polly was different. She stopped practising as a barrister and instead began advocating for an international law that would make business executives and governments criminally liable for the environmental harm they cause. Environmentalist George Monbiot, writing in the *Guardian*, describes the impact this might have: 'It would radically shift the balance of power, forcing anyone contemplating large-scale vandalism to ask themselves: "Will I end up in the international criminal court for this?"'

In the summer of 2020, ecocide was all I could think about when I heard that a large vessel had crashed into a reef off Mauritius after supposedly going too close to the shore so that the captain and crew could get a Wi-Fi signal. The MV *Wakashio* was a Japanese bulk carrier holding over 4,000 tonnes of fuel. For twelve days the ship sat jammed into the reef, with the locals slowly becoming aware of this gargantuan foreign object invading their pristine waters. Then, on 6 August, its hull began to splinter and split apart, oozing and spitting millions of gallons of engine oil into the reef. The oil spread outwards like some sickly virus and began to enter the local wildlife refuge, which was a sanctuary for species endemic to Mauritius that were on the brink of extinction and needed a safe space free from humans where they could luxuriate in the turquoise waters. The strange thing was that although this was one of the most catastrophic Indian Ocean oil spills ever – a lethal injection into the heart of the ecosystem, smothering its fatal liquid over the abundance of life in the reef – we still sat back and called it an accident. When does the word 'accident' become inappropriate? When will we realize that such 'accidents' are actually becoming inevitable events due to the way we live our lives? When will we notice that unless someone is held accountable, this consistent pattern of catastrophes will continue?

I wanted to speak to some young people who are today continuing Polly's work of accountability.

On 29 October 2012, Vic Barrett was hiding behind his mum's bed. The howling gale outside seemed almost human, out to get him. In a way, he was right, I suppose. Superstorm Sandy *was* less natural than most storms up until that point, and more malicious. It tore through the world like a human bulldozing through a rainforest, with no care for the life crushed beneath its path. Vic was twelve years old at the time, and spent days hunkered down in his mum's bedroom without power 'as the storm raged around us . . . waiting for one of the many trees surrounding my house to come crashing down on us, just like I'd seen happen on the TV'.

Vic is a first-generation Garifuna American. His family are an Afro-indigenous community originally from the island of St Vincent in the Caribbean, who moved to the east coast of central America (Honduras and Belize) in the eighteenth and nineteenth centuries when British colonial powers dominated their homeland. His mum then moved to New York, as she thought she could create a safer, more stable upbringing for Vic there.

On 29 October, Vic was sent home from school. The New York Stock Exchange shut down. The city's bridges were sealed off, the subways and airports closed. Hundreds of thousands of people streamed out of the city, but at 5.55 p.m., Mayor Michael Bloomberg warned that Superstorm Sandy was the 'storm of the century', and that the time to evacuate was over.

The water kept coming, turning the city into a tangle of rushing brown rivers and rubble. The streets became saturated with the consequences of human action. It is in moments like these, when humans are being pounded by the foam and surge of nature, that we realize just how deeply delusional our 'dominance' is. This is a world just one degree warmer than pre-industrial levels. Dare to cast your mind to a world where we don't change, setting the stage for a potential five degrees of warming by the end of this century.

Climate change didn't cause Sandy, but it certainly created the circumstances that fuelled it. The storm surge that slammed into the coastline from Massachusetts to Maryland was the most destructive part, and it was global-warming-related sea-level rise that provided the surge with a higher launching pad, making it much worse than it otherwise would have been.

Eventually the iron-dark sky lightened and New York was no longer shrouded by towering clouds but instead illuminated by the receding glow of distant lightning. Much of the city began to throb back to life, with people sweeping up leaves and rubble and wading through muddied streets. Meanwhile, Vic and twenty other young people dotted across the affected areas were hatching a plan. They began creating a novel legal theory – that a safe climate should be a civil right, and that through policies like leasing public lands for coal mining, the government was violating this right.

In 2015, they filed their constitutional climate lawsuit, Juliana v. United States, against the US government. Their complaint asserted that 'through the government's affirmative actions of pursuing policies that exacerbate climate change, it has violated the youngest generation's constitutional rights to life, liberty, and property, as well as failed to protect essential public trust resources'. The twenty-one young people wanted a conclusion where the government would agree to a plan to phase out fossil fuels and pull greenhouse gases back out of the air.

Vic explained that 'the executive branch of the United States government – a government that, in theory, was created to protect us – has known that burning fossil fuels will cause climate chaos for more than fifty years. Instead of working to curb emissions, it has colluded with corporate interests – actively enshrining the fossil fuel energy system into our nation's policies. Because of the actions taken by the United States government and the fossil fuel industry, my generation has never known a world free from the impacts of climate change.'

I scrolled through articles about the lawsuit, already half aware of these young people yet still feeling a surge of awe – as I often do after speaking to other young campaigners. I pored over their individual stories and the processes they went through. I tracked the case as it wended its way through the federal courts, which it did for years. It survived several motions to dismiss, and also intervention by the US Supreme Court. Then, suddenly, on 17 January 2020, the lawsuit was dismissed. Just like that.

The twenty-one plaintiffs refused to give up there. Together with their lawyers, they stated their intent to 'appeal this dismissal to the full Ninth Circuit sitting en banc'. Essentially, they're now pushing for another hearing, with several judges in attendance rather than just the one.

In activism, there is no such thing as an action that simply fails. Just like an ecosystem, each species, each act has webs of causes and effect that ripple out, reverberating and shifting the stories of unwritten histories. The Juliana v. US plaintiffs are still fighting for their cause, which may or may not be accepted by the court. However, hundreds of other people have been inspired by what they're doing, and youth-led lawsuits are awash around the world. Young people are claiming a right to a clean environment, and many of them are winning.

A group in the Netherlands sued their government, which led to a decision to curb carbon emissions by 25 per cent in one year. In Colombia, twenty-five young people won their lawsuit against the government for failing to protect the Colombian Amazon rainforest. The court agreed that deforestation violated the rights of the natural world, and the government was ordered to reduce deforestation to net zero by 2020. Simultaneously, in Pakistan, a seven-year-old girl gained the right, for the first time ever, for a minor to sue in court. Her action was successful, and resulted in a climate change commission to monitor the government's progress and hold them to account on their pledges.

There have been many methods used to try and force governments to act in favour of the planet, stemming from all ages, ideologies and locations: petitions, boycotts, hunger strikes, civil disobedience, marches and whistle-blowing. On 14 April 2018, a sixty-year-old man ambled into Prospect Park in Brooklyn, soaked himself in gasoline and set himself on fire. A note found a few metres from his body read 'I am David Buckel and I just killed myself by fire as a protest suicide.' In a longer typed letter he had earlier sent to the media, David elaborated on how his early death by fossil fuels was symbolic of what we as a species are doing to the planet.

Fortunately, the twenty-one young people involved in the Juliana v. United States lawsuit didn't take such drastic measures, but people *are* becoming increasingly desperate. On Christmas Eve 1968, *Apollo 8* was looping around the moon. On its fourth lap, the crew saw our planet in all of its beautiful insignificance – a bright blue and white dot suspended in the endless black space. The photograph they took is known as *Earthrise*. When people saw this photo, they began to see the world as a singular, highly fragile oasis. On that pinprick has lived everyone we've ever known. All the beauty, art, poetry, music to emerge from our species came into existence there. Now we are going through another Earthrise. The world is finally awakening to the reality of how finite our planet is, and many may feel compelled to go to more drastic lengths to protect it.

Manifesto for change

- Talk about it. This is one of the most important things you can do. Speaking up about environmental issues will likely start a chain reaction and hopefully cause many others, young and old, to speak out and take action too.
- Reduce your greenhouse gas emissions. See Chapter 2 for more.
- Make your home as energy-efficient as possible. For example, replace halogen bulbs with more energy-efficient LED bulbs, wash clothes at a lower temperature, turn your thermostat down, insulate your home as much as possible (walls and roof), fit doors and windows to avoid draughts, reduce your use of home appliances and install a smart meter.
- Use your power as a consumer. Always consider the carbon footprint of the products you buy. What are the air miles of your food? Can you buy

more local and seasonal products? Think about the environmental impact of practices like deforestation, used to produce unsustainable palm oil; avoid products containing it if and where possible.

- Support environmentally responsible and sustainable companies such as Beyond Meat, Klean Kanteen and Abel & Cole, as well as local independent businesses. Do a bit of research into companies' credentials and then repeatedly return to them and recommend them to friends.
- Join protests. The power of citizens coming together through collective action cannot be underestimated. It's more than just a means to change a decision-maker's mind. Sometimes, being in a throng of change-makers and feeling the energy, passion and hope of a crowd helps to recharge you and push you to continue fighting. The school climate strikes and the Extinction Rebellion protests have both had measurable impacts, with governments around the world making commitments that were unimaginable before. It's not enough yet, but they're steps in the right direction.
- Join campaign groups and organizations such as 350.org, Greenpeace and Earthjustice calling for governments to:
    - stop investing in and subsidizing the fossil fuel industry. This money could be used to support

healthcare, natural restoration, green infrastructure projects, renewable energy and other essential priorities. This will allow the money to be channelled to more sustainable ventures, whilst also supporting global decarbonization and creating an international market for renewable energy technology

- diversify their supply of energy, as this weakens the dominance of autocratic petrostates, such as Russia, whose supply of fuel many countries are highly reliant on
- redesign energy markets so that there is a bigger safety buffer and energy suppliers hold more reserves, just like banks carry capital, to enable them to absorb shortages and deal with the vulnerable state of renewable power.

Chapter 5

Our Disconnect from Nature

Reconnecting young people with the natural world

Ironically, I watched the report on TV whilst quarantining in a dingy Parisian hotel room. Through the French doors came a gentle breeze that stirred the curtains and carried up snatches of excited conversation from the street several floors below. Some wild violets I had been given at the start of the week had turned putrid and murky in their water, and so my saving grace was the rectangle of sky I could see through the panelled door. I could watch birds dipping their wings in the orange of the rising and setting sun. It was only seven days, but by the end of it I felt an insatiable craving and desire to be back in nature again.

During one of the long, hot days that I was boxed into that small room watching vacuous stuff on the TV, a report began on something called 'hikikomori'. This is defined as 'individuals who refuse to leave their parents' house, do not go to work or school and isolate themselves away from society and family in a single room for a period exceeding six months'. Some, though, such as a young man named Jinsei in the report I watched, will isolate themselves for decades at a time without setting foot outside their apartment once. As I continued watching, the scale of the crisis shocked me. There are over half a million hikikomori in Japan, and the number is ever increasing. I flicked off the TV and lay in my cramped room, trying to imagine spending another twenty years in there. Then it struck me. What the French TV channel were calling a crisis for Japanese society was not just a Japanese problem after all. It was a problem caused by modernization and urbanization, and it was a global issue.

No one knows anything, really. Everything we think we know (other than instinct) is borrowed knowledge, fed to us by the spoon of culture, society, family or friends. A human born into the world is like a lump of clay that has the potential to be moulded into someone capable of living sustainably on this planet, but also into someone who will risk everything, including the planet itself, for money, fame or a new car.

So that's where I'll begin – at what I believe is the root cause of this crisis we face: the way in which much of the Western world moulds young people into nature-fearing, indoor-loving adults. It's time to talk about something called nature deficit disorder.

Nature deficit disorder, a term originally coined by author Richard Louv, is an environmental crisis in its own right. Three quarters of UK children spend less time outside than prison inmates, according to a study by the National Trust, and less time than any other country in the world. This is due to an attraction to screens, the disappearance of green spaces and the ever-growing threat of stranger danger. Like Louv, I'm wondering where the future generation of environmentalists will come from. The burgeoning enchantment with our phones, TVs and iPads seems to be swallowing up and spitting out our once-amiable relationship with the natural world. It is a persistent, pernicious cycle driven by generations of nature deprivation, one that is churning out more and more nature-deprived children. There is, however, a solution to break the cycle, and it lies in ancient pools of knowledge.

In the Western world, as our human dominance of nature has grown, we have become more isolated. Philosophers call this state of disconnection 'species loneliness'. The human species is like the bully in the playground who watches as all the other children play together nicely, whilst we sit estranged in the corner bitter and regretful. You see, we've reached the top and realized it's very lonely up here. We tend to view nature as something apart and completely separate from ourselves. Our social construction of trees, birds or flowers is mechanistic and detached, and we are quick to deem any other cultural constructions primitive. When we hear that some indigenous cultures firmly reject the notion of 'nature' in opposition to humans, it feels like our deeply held assumptions are being challenged. It's uncomfortable.

When you take a moment to jump out of our paradigm and really examine another, though, it begins to make sense. In the Western world, the word 'nature' is a euphemism for anything not human-made, often something wild, untamed and dirty. We have posited nature as the antagonist of civilization. By removing ourselves, cities, cars and buildings from nature, we create a dichotomy that suggests we can somehow exist in a world isolated from, or even without it. We can't. Nature means nothing unless it means everything.

I've always loved spending time in nature, but when I was younger, I used to have an intangible revulsion of fields. I didn't understand why, but on family walks I would feel a perceptible darkening of my mood as we stepped out from the forest into a field. Suddenly my wellies were leaden, and I would become bored and moody as we traversed this immensity of close-cropped, controlled nature. I liked disorderliness before I knew what it was, as many of us do when we're young. Even when we grow up, that love of the wild and untamed sometimes surfaces. Last weekend, a seemingly solid tree on a road near my house fell down during a storm and became almost a tourist attraction, as people took photos of nature blocking humans in their cars. When the fabric of normality is ruptured, we feel a little thrill, a little excitement.

In February 2019, I learnt that there was another way to think about the natural world in relation to ourselves, another way to live within it and alongside it. We were finishing off six months of filming for the environmental documentary *ANIMAL*, by Cyril Dion. Our last shoot was in Costa Rica; a good place to end because of its ecological success. The country abolished their army in 1948, committed to carbon neutrality before almost all other countries, and is now one of the most biologically diverse areas of natural beauty in the world. They have a world-class health system and are rated the happiest and most sustainable nation on earth (according to the Happy Planet Index). Costa Ricans say '*pura vida*' ('pure life') to each other just as much as Brits talk about tea and the weather.

We were there to film members of the Térraba tribe.

Although we were on a strict filming schedule, we were told not to expect punctuality. In the UK, 'if you're right on time, you're late'. In Costa Rica, you follow 'tico time', meaning people come late, later or never. When the tribe members rolled up the dirt road to the agreed location, the sun was already glowing red and the film crew were sprawled on benches and in the thin shadows stretching out from the side of the cars. The heavy, syrup-like air had made us sullen and sweaty, but the Térraba members tumbled out from their vehicles in a haze of face masks and feathers, bellowing a collective *pura vida* before the car doors were even shut. As we travelled with them, it became clear that we were different in more ways than just punctuality.

Twenty days after I returned from meeting the Térraba people, when the thrumming forests of Costa Rica were a mere memory for me, I was sent an article from the *Guardian* about how a land activist had been killed by an armed mob whilst trying to reclaim his ancestral lands. It was a tragic realization when I saw that the man killed was from the very tribe who had welcomed us so warmly around their fire and into their home just weeks before. They were asking for their stolen place to be returned and were met with death instead.

In modern society, our sense of place has been lost over time. In fact, I would go so far as to say that much of the anxiety and moral uncertainty we experience today is rooted in our loss of a deep sense of connection with the natural world and physical places that help to define us. We spend so much time lurking in virtual spaces and communities, but cyberspace, it turns out, is a poor alternative for the real thing.

The sense of place we do have is very egocentric. You might remember where you had your first kiss, where you fell off your bike or won a football match. You will feel a powerful nostalgia in your childhood neighbourhood; the landscape around you will be pockmarked by memories of your most numinous, clarifying, devastating moments. What I sensed with the Térraba tribe was a much more impersonal, yet paradoxically intimate sense of place. For them, the land does not revolve around the individual; rather, the individual embeds themself in the nature of the land. They understand the calls of different birds, the species of trees, the texture of the earth, the patterns of the weather. Each of these things is as much a part of their survival as their heart and lungs. They see their place and environment as an extension of themselves rather than some separate entity.

If the Térraba can fight for their forests and green spaces, why have many of us not tried to do the same? 'What green spaces?' you might ask. In western Europe, we have lost more than half of our forests. Nature deficit disorder is more than just a few too many hours in front of a screen. It also concerns the fact that we evolved as biological beings in a biosphere, yet now inhabit a technosphere. It describes the rigid demarcations we have created in the Western world between humans and nature, ourselves and others, ourselves and our 'place'. We see those we must protect as the people immediately around us, and our home as only the four walls within which we sleep.

As a product of this society, I inevitably carry within me some of the very structures I seek to question and overthrow, so let's put this upbringing under the microscope and examine the parts. It's deeply unnatural, perhaps even impossible, for many of us to be objective without being influenced by culture and society, family and friends. We are, after all, the only animal that feels we must solve our existence and give ourselves a purpose. I would argue that in this age, our purpose is to reconcile our relationship with the natural world. Let's first try to define this relationship.

'Wilderness' has come to be an antonym to 'civilization'. We tend to see being 'civilized' as a desirable trait, so where does that put 'wild'? Oddly enough, despite the common revulsion to mud and the most elemental aspects of nature, we are fascinated by the wilderness. The story of the 'wild child' captures the imaginations of everyone. Whether it's Mowgli in *The Jungle Book*, Romulus and Remus, who suckled from a wolf, the exposure of Paris on the slopes of Mount Ida, or Tarzan, there is something undeniably attractive to us all about those who are nurtured by (non-human) animal kind.

In 1996, the fabric of Russian society was fraying and damaged after the fall of communism. Poverty was rife, and many children were left on the cold streets, becoming an invisible part of the landscape. Ivan Mishukov lived in an apartment permeated with the acidic stench of alcohol, and his mother was too weak from abuse to care for him. So at just four years old, he wandered out onto the streets with no belongings, into the wild. He begged for scraps of food, and would tear up his rations and share them with a local pack of dogs. Gradually he became ensconced in the pack. He would walk on all fours and sniff his food before he ate it. He was, by society's standards, uncivilized, *wild*. We think of nature as hostile, but Ivan had left the cruelty of the civilized world and moved into a kind reconciliation with the animal world. He had found a sense of belonging in the wilderness he had never experienced before.

Three times the police tried to wrest Ivan from the grip of 'savagery', but the snarling boy was always defended by the dogs. Only when they separated him from the pack one day did they manage to capture him. He was sent to an institution, and it's said that for a while, the dog pack waited in vain for him at the gates.

I'm telling you this story because it blurs the boundaries between 'wild' and 'civilized', between 'us' and 'them'. Holding animals and the natural world at a great distance from ourselves allows us the freedom to do with them what we wish; to harm them. But if a boy can become dog-like so easily, perhaps we need to redefine what it means to be human. If this boundary is so easily blurred, it is the fault of the culture we're brought up in that we're so separated and disconnected from nature; we are not born like that.

George Monbiot has described what he calls 'shifting baseline syndrome'. Each successive generation sees the wildlife of their own time as the norm, failing to realize that some populations have plummeted or species that weren't there before have rocketed. As a result, people assume the degraded environment around them is the norm. Young people don't know that they should be seeing flocks of starlings or house sparrows, because they've never experienced their presence. The scientific data may suggest we're in a crisis, but it's hard for people to believe the facts when everything around them feels perfectly mundane, just as it always has been (especially if you're a teenager).

So, we've established two main things that contribute towards ecologically impoverished childhoods. Firstly, the dominant cultural narrative that puts humans on a pedestal as somehow above and separate from nature, and secondly, the belief that everything is fine and our meagre green spaces are just how they always have been. Nurturing a love of the planet in young people is the easiest way to combat this. Many are going to have to make the effort themselves, lace up their boots, venture outside and fight for that love until it becomes something visceral, until the flap of a bird's wing or the rustle of a mysterious unknown creature in the bush makes their pulse quicken.

I want to tell you now about someone who *has* changed, because radical change is always possible. Someone who grew up one way and then became the antithesis of his past. His name is Daniel Valencia.

The best way to get another person to change is not to push them in any specific direction. Rather, walk ahead in the desired direction yourself and ask them to come with you. In this spirit, I present to you, like an acorn, Daniel's story. You can decide to leave this acorn on the ground, at your own peril, or instead nurture it, and one day you may be lucky enough to sit under its protective shade.

My path and Daniel's have never crossed physically, but they intertwine in the work we do. In 2019, I helped to set up Reserva, an organization completely driven by young people that protects areas of land around the world. Our first reserve was in Ecuador, in a humming bounty of life called the Chocó rainforest. I say humming bounty because there are so many species jostling for space in the Chocó, but only 2 per cent of the area remains due to rampant deforestation as a result of mining and intensive agriculture. So the hum is quieting and the bounty is being carelessly hacked away at by human hands. The prospect of paradise lost is heart-wrenching for me, but for Daniel, the Chocó is more than a remote paradise; it is his home.

The story originally begins thousands of years ago, when humans first moved into the Chocó, or perhaps in the 1500s, when the first Spanish colonists wedged the hull of their ship into Ecuadorian soil. But I want to focus on Daniel's part.

Back when he was a young boy, Daniel was a hunter. Not in the trophy-seeking, Range Rover and gilet type of way we might imagine, though. He was taught to treasure the thrill of it, and learnt the skill of searching for tracks amidst the foliage. He felt a strange intimacy and respect for nature, as he had to try and think like the animal, and even be the animal, in order to understand its behaviour and track it down.

He hunted anything and everything he could: monkeys, olinguitos, cuchuchos, foxes, armadillos, guantas and guatines; hummingbirds, cockerels, hawks, tanagers and tyranids. He explained that it wasn't done for subsistence, rather 'for harm or taste, at that time due to ignorance'. It was about the anticipation, the bated breath, the climax of the kill and the trembling hands claiming the innocent animal as your own as you watched the life pour out of it. It was primal adrenaline-fuelled sadism; it gave him a sense of power. In that fleeting moment, he would forget about everything and think about nothing.

Anything that allows humans to do this is addictive in a way. For some it comes in the form of drugs, sex, food, sport or art, but for Daniel, it was hunting. To me, pursuing a wild animal and killing it instantaneously seems less morally offensive than visiting a drive-through fast-food chain and picking up a bucket of meat from an animal that has spent its life in a cramped cage. For Daniel, however, it was wrong enough that very soon he began to question his actions.

The moment he turned from a hunter to a conservationist was distinct. When he was around fifteen, he was invited to do some birdwatching in Colombia. When he and his companions arrived at the site, they sat waiting and watching, soon to be drenched by the heavy, onrushing clouds of the rainforest. Then, in the last shards of light, right in front of them appeared a little *Manacus manacus*, a bird that looks like a speckled old chimney sweep with a flowing white beard and a tight black cap. As it snapped its wings impatiently and hopped from branch to branch, Daniel adjusted his grip, lined up the shot and pressed.

When he got home, he laid out his prize on the table and admired what he'd taken. But this time it was all ink and memories, not blood and feathers. He had decided to shoot photographs rather than animals, and he never went back to his old ways. He had realized that admiring nature didn't have to mean freezing it in the timeless grip of death so that he could marvel at a creature on his wall, rigid and enduring. Rather, he could capture the moment on a camera and allow the animal to continue existing. There would no longer be the thrill of the pursuit, or the explosive excitement of the kill, but this new direction for Daniel was something much more.

Now, aged twenty, Daniel is a park guard in Carchi province, Ecuador, where he works to monitor the Reserva site's population of critically endangered brown-headed spider monkeys. Since the day the *Manacus manacus* burst into the last shards of sunlight, he has become one of Ecuador's top birders. He has broken the Carchi province record for the highest species count, and is now learning to monitor not only the monkeys, but orchids, reptiles and amphibians too. He went from predator to protector, from dominator to steward, and that's what we as a species must all do.

Daniel speaks Spanish with flecks of English, and I speak English with flecks of Spanish, so he sent me his story in a document, which I then translated and pored over. I was like a pathologist dissecting and analysing, searching desperately for the answer I knew was contained in what lay in front of me. What could change a man so completely, and how could we do that for everyone? Ultimately, although we constantly use phrases such as 'fighting against the climate crisis' or 'fighting extinction', we're really just fighting against ourselves. There's no great force or superior race coming down to challenge us; everything begins with changing hearts and minds. Whether this is a relief or a terror, only time will tell.

Around the same time that I realized we must completely change how we see and relate to nature, I also stumbled across a film called *Captain Fantastic*. It tells the story of Ben and Leslie, a couple who take their children to live in the woods after they become disillusioned with American capitalist society. The film begins with one of the children disembowelling a deer, before scaling a sheer rock face, racing through undergrowth, climbing trees, and foraging for food. It continues in this way, showing them as feral and wild, not materialistic or addicted to processed food and video games.

Captain Fantastic might be a fictional story, but it does provide us with valuable lessons and insight if we are to begin truly 'rewilding' young people and reconnecting them with nature. I wanted to understand how we can bridge the gap between, on the one hand, being so ensconced in modern life that we disregard nature, and, on the other hand, becoming so disillusioned with modern life that we remove ourselves from it almost entirely. There's a fine line between the two and it involves compromise, but we need to find it.

I spoke to the non-fiction version of *Captain Fantastic*, a young British family who had uprooted themselves and moved to the wilds of New Zealand. I wanted to hear the children's stories, but they really begin when Lucy Aitkenread, their mother, was a young girl.

I first spoke to Lucy in early October 2020, and now, in December, I am finally sitting down to write her story. I've been turning her words and wisdom over in my mind like hot stones, unable to settle them. Knowledge about the failure of our educational system has always been quivering on the horizon of my consciousness, but has been just out of reach. When you're ensconced in the system, it is all you know. The dark winter commutes to school on the bus, the mantra of the morning register, the cracking-open of a new textbook at the start of term, the teachers you like and those you don't, the cliques, gossip and unspoken rules of adolescence. And because it's all you know, it's hard to have any heightened sensitivity to how different the system *could* be. When you're old enough for your opinions to be taken seriously, you've already been through it and most likely are too busy being consumed by the stresses of adulthood to worry about changing it.

Lucy, who was born in Britain, met Tim whilst studying in New Zealand. They married in 2005 and moved to London, where they had their children, Ramona and Juno. Gradually, the more their lives conformed to the conventional standards of what normality and happiness are supposed to be, the unhappier they felt. The mortgage, the rat race, the constant technology, and the fact that they had to thrust their young children into care whilst they slogged off to work every morning – none of it was conducive to what they'd hoped life would be like. So they did what they felt they must, and traded the comfort of their home and life in London for the chance to nestle themselves in the native rainforest of New Zealand. They moved to a yurt – a large tent traditionally used by nomads in Mongolia and Siberia – in the remote wilderness near the Karangahake Gorge, and from the very land itself they built their home around them, like birds carefully constructing a nest. They installed solar panels for electricity, a wood-burning stove for heat in winter, and a vegetable garden for self-sufficiency.

At this point in our call, Lucy flips her camera around and shows me a herd of free-range guinea pigs swaggering past her like a group of pub-goers leaving the bar on a Friday night. They stop, bristle nonchalantly, look at Lucy, wrench out fat mouthfuls of grass and then swagger on again. They're symbolic, in a way, of the freedom Lucy tries to embody and create within her home and family. She's free, her kids are free and her guinea pigs are free. However, this desire to connect with nature, to leave behind completely much of modernity and seek something else, isn't inherent; it is a product of Lucy's early years at school.

As a teenager, she was caught in a vicious cycle of bunking off school and then, as punishment, being confined to a small cupboard called 'isolation' for several hours at a time. This caused her to hate school so much that she'd bunk off again. 'The good times,' she recollects, 'were when I wasn't caught, and I'd get home and shake sand from my shoes and flick leaves from my hair.'

Lucy's teenage years fulfilled every catastrophic adolescent archetype out there. It was like a kind of hormonal thunderclap, where she veered from pious Christian fundamentalism to pure delinquency and back again. She spent much of her time weaving between arcades and alleys, bleaching her hair with chemicals from the science lab and playing Snake on her Nokia. Every instinct told her to remove herself from the coercive, manipulative school she went to, so she did. She'd 'skip off the bus – sometimes with friends, sometimes alone, and roll around town. School tie shoved in pocket, Miss Selfridge lipstick on and eyebrows somewhere way back in the early nineties.'

This is the story of a teen rejecting the education system, rebelling against authority and trying to find some shred of identity in the chaos of adolescence, but it's also the antithesis of who Lucy is now. She spends her days taking her children to ancient woodlands and windswept coasts, allowing wonder and natural curiosity to dictate their lives.

It's very rare for an idea to embrace you like an arm and lead you away from everything you thought you understood. All the previous attempts I've seen to try and derail humanity's sprint away from nature have been jabs at 'the youth of today'. 'Too much screen time,' they say. 'Oh, those damned sedentary snowflakes. They are all becoming obese. Not enough fresh air . . . Their eyes will turn square from staring at screens all day. Such a generation of narcissists that the selfie is the zeitgeist of the era. They are so enthralled with their spheres of connection that they are more disconnected than ever.' This is the narrative firmly pinned to my generation, and I bought into it. I distinctly remember my childhood brimming with freedom, but slowly this memory has been picked apart by the assumptions about young people until all that is left in my mind is the bright glow of a screen and the ghostly little Snapchat figure that haunted me whenever I went on my phone.

I was convinced the devices were to blame for my generation's shocking alienation from nature, but Lucy says otherwise. The reason is systemic. It's deeply embedded in the very material of society, and it's the way in which we bring up children that has to change. You are either a child embedded (or soon to be embedded) in this system, or an adult who is upholding or opposing it. Lucy's life is the extreme version of opposing the system, but as you should have concluded by now, we can learn lessons outside of school, and the lifestyle of Lucy and her children holds lessons relevant to us all.

It all began for Lucy when her young family took a trip to the Black Forest in Germany. There, whilst they searched for somewhere to educate their children, they visited a *Waldkindergarten*, or waldkita, a German forest nursery. The weather was drab, the sky grey and the day gloomy, but the children's eyes were bright, their cheeks flushed and their hair wild as candyfloss. Together the huddle of kids boarded a bus and travelled to an 84-acre park. The pedagogical philosophy of waldkitas discourages toys, so they arrived at the park empty-handed. Without toys with pre-defined purposes, their imaginations were unboxed, set free. They ran amok, out of sight even, some clambering up trees, crouching in the boughs and manoeuvring themselves through the tangle of limbs with ease. Others slid unsupervised across a frozen pond, peals of laughter echoing over the ice. But whenever the children heard the 'cuckoo' sound from an adult, they dutifully returned and arranged the blackberries and other fruits they'd collected on a platter to share with everyone.

'That was the thing that made everything click into place for me,' said Lucy. She and Tim decided then and there to remove their children from their London primary schools and move them to the wilds of New Zealand to begin the process of 'unschooling'.

Whatever you think of education, the city and our disconnect from nature, we as a species are undoubtedly enduring an epidemic of depression. This is a time of collective existential crisis – we communicate in perfunctory emojis to tell people how we feel, we drive from place to place because the roads are too polluted to cycle on. 'Sometimes in the city, I can almost hear the hiss of steam as people's minds vacate their bodies, unable to stay put in the moment. Analysing, planning, rushing, ticking off, typing, gathering stats, brainstorming, texting,' Lucy says.

That's the problem, you see, it begins at school. We institutionalize nature deficit disorder in children who are barely old enough to say the word 'butterfly'. Not just through the way young people are physically removed from the natural world, but also in how their imaginations are ensnared by swathes of sums. They are taught not to gaze out of the window, not to leave class to go to the toilet, not to touch mud or climb trees, not to embody the wilder side of their idealistic, joyful youth. Research from Cambridge University shows that 'children who start reading later (at seven as opposed to five) quickly catch up to their peers and by the age of eleven show better text comprehension and more positive reading attitudes than their early learning peers'.

I'm almost out of formal education now, so this is not the 'save me from SATS' type call for help you might assume. It isn't even about home schooling or forest kindergartens; it's about a collective move away from our industrial model of education. We manufacture brains like we manufacture computers, based on conformity and batching people together. The solution is, like with our agricultural systems, using a less intense, more organic process. As educationalist Ken Robinson said in his renowned TED talk, 'We cannot predict the outcome of human development. All you can do, like a farmer, is create the conditions under which they will begin to flourish.'

As we finish up on our call together, Lucy's audio crackles as she runs suddenly through the dewy morning grass back to the house, the camera jostling wildly. Ramona is on the floor, crying, one leg ensconced in a cast and her crutches piled on top of her. Lucy holds her to her chest and strokes her hair, explaining to me that in her ten years of life, Ramona has broken every limb. Her sobs subside into whimpers, which soon blossom into giggles as she gathers her crutches and hops into the moist breath of the golden morning. At nearly eighteen, I haven't broken any bones. It used to be a badge of honour for me, but seeing Ramona hop and weave through the thick glades and grasslands, I feel like a penned rabbit watching wild hares frolic outside my cage.

I have one more question for Lucy. What about privilege? What about the families living in the debris of wasted potential and opportunity? The ones with their hands to their mouths, their foot barely in the door of life, fighting under the weight of just trying to make sure their children survive? Not everyone is in a position to be able to up and move their family across the world. A parent struggling to feed their child will probably have other concerns and priorities that don't necessarily revolve around that child's exposure to nature.

'I could feel guilty about it,' Lucy says. 'I could be paralysed by shame, but instead I'm using my privilege as a fuel to create this opportunity for my family, and now an opportunity to help parents around the world do the same. Injustice, trauma, inequality, these are all fuel for our tanks. I've worked with BAME parents in the US who are using unschooling as they're opposed to the systemic racism within the school system. They are turning the injustice into the fuel itself. Lots of the people I know doing this are powered by the hardships they've experienced, and a desire to carve out another way of life.'

When I finish speaking to Lucy, the wind is rising and the sky is a bright, cold blue. I plunge out into the autumn day and cycle wildly to my favourite spot, my favourite tree by the river. The bark sags like the skin on an elephant's leg, and I use the calloused grooves to shimmy myself up into the limbs. It is nice to be in the tree, to be in the dancing branches shivering under the strain of the wind, watching the birds hop and whir. I also enjoy the moments when people pass on the path just below. I sit there knowing that as long as there are trees on this planet, every child (and every adult) should sit in the boughs and see the world stretching out beneath them. It seems that problems and worries can't climb trees; they can never reach you up there.

Manifesto for change

- Community. Know your neighbours and together invest in community playing spaces for kids. Create play groups where parents take it in turns to keep an eye on the children in local parks. Same 71 per cent of adults used to play outside, in comparison to only 20 per cent of youth now (a poll by Play England).
- Help make your town/city greener. Do this by pushing for sustainable urban planning with your local council. With a local group you can drive redevelopment in your area: e.g. tree planting, walkable neighbourhoods, increased public transport so there are fewer cars on the roads, more diverse and natural parks.
- Read Richard Louv's book *The Last Child in the Woods*. Make sure to share and amplify the key issues discussed. Be the one to start that conversation and create ripples of change.
- Challenge the restrictions that prevent or prohibit natural play in your area. Whether they're imposed by the council, or just by parents and authority figures, don't be afraid to question them. A survey commissioned by Playday found

that 77 per cent of children would like more opportunities to take risks such as climbing trees whilst playing, and that this less tame method of play makes over 90 per cent of children feel happiness and satisfaction. UK statistics showed that in 2007, children were nearly three times as likely to be admitted to hospital after falling out of bed than falling from a tree.

- Collaboratively renovate decaying or derelict areas into natural spaces or 'wild zones'. Developers will often neglect slices of land that are not large enough to be parks or playing fields, but you can persuade the local council to create islands of wilderness, even if they are only pockets of green for children to enjoy on the way home from school.
- Seek out natural play designers. These specialize in creating living landscapes and play areas for young people in your community.
- If you're a student or teacher, create and promote nature clubs and activities at your school. If you're a student, it may feel daunting, but trust me, many others will be wishing they'd thought of it and want to join you.
- Break down the barriers by engaging in dialogue with people of different cultures, religions and demographics. This will help to ensure that natural spaces feel inclusive and accessible for everyone. You can even organize face-to-face conferences locally or online internationally to

discuss the issue of nature deficit disorder and begin exposing the challenges and hesitations that face different groups when trying to access nature.

- Contact botanical gardens, parks, churches, community centres and natural history museums and ask them to host discussion groups. They might be able to put on nature talks or workshops for young people.
- Reach out to people digitally. You could try joining apps like Meetup or Nextdoor to meet like-minded young nature enthusiasts and organize/host nature-oriented activities with them.
- Rewild yourself, your garden or a local allotment. Introduce native flora and fauna. Go camping. Go outside every day. Take night walks. Plant a tree. Go seasonal foraging. Buy natural history field guides. Clean a beach. Go wild swimming. Try star-gazing. Learn the constellations. Listen to the dawn chorus. Watch the sunset. Roll down a hill. Press flowers. Plant your own fruits and vegetables, and invite others to join you . . .

Chapter 6

Women and Water Insecurity

The connection between women and nature

I love wild swimming. It is an escape from squeezing breathlessly through the city throngs. In the water it is a different world, softer than the land. You see different life forms, or the same forms but in a different way. The milky bellies of swallows diving into the lake, the feathered rainbow explosion of a kingfisher launching off a branch, and the aquatic life darting behind the reeds. Many cultures see water as a symbol of purity and life. Human beings are made up of roughly 60 per cent water. You've probably drunk it today, showered in it (possibly), run through it as you struggled with an unforgiving umbrella (most likely for the British among us).

Seventy-one per cent of the earth is covered in water, and yet only about 0.007 per cent of that is available to fuel and feed people. That, combined with our wasteful ways, means that within the next decade, global water demand is expected to outstrip supply by 40 per cent. We won't all wake up one day to a thirsty planet; some areas will watch as their resources dwindle down to drips, others will find pools they'll never swim in. In fact, the process has already begun . . .

Paul Mutuku grew up in Kiteng-ei, a village of only 500 people. It is so small and remote that it has almost eluded the scrutiny of the virtual modern world. I stumbled through the recesses of the internet searching for any scrap of information on this village, but came up empty-handed, except for one study published in 2009 that examined occurrences of 'intestinal polyparasitism in a rural Kenyan community'. Although life was not easy, Paul said he 'seized every moment to go to the forest and collect firewood, graze cattle and goats and play with my peers. Things were just beautiful. We lived alongside bush babies, cheetahs, wild cats, chameleons, geckos, antelopes, hares, birds, among others.'

On a pale, cheerless British day, I listened as he spoke tenderly of his childhood, depicting a poetic utopia. In that childhood he was encouraged to grow outwards into the world, not downwards into a phone screen. He would wake up to the sounds of children playing, birds chirping, communal cooking. He and his three brothers would weave clothes together to create a football. 'It taught us creativity and innovation,' he reminisces. 'We would skip and sing with the girls . . . it taught us how to deal with the opposite sex. Before that we would hide from the girls in thickets of trees. It was just organic. As a kid, I accompanied my grandfather as he went to plant trees, and I remember how fun it was to play in the muddy black soil during the planting activities. I also helped him prepare nursery beds for trees. Those are the roots that moulded and define the person I am.'

It was those roots too, Paul says, that shaped his deep understanding of the natural world. 'Nowadays we hear young people saying water comes from bottles, or food comes from the supermarket, but for us in the village we had to learn that water comes from the trees and the springs, and if you fail to protect the diversity around that tree or the rocks around that spring, then you're not going to drink.'

It is then that the cracks begin snaking into the story. Paul's words are not at all self-pitying, but you can tell his history has been charged with pain. 'I am from a single-parenthood. Usually if your mother was out you would be left with your father, or vice versa, but we were left with the reality that our mum was forced to fend for us, and that meant being left alone.' He hesitates for a moment and looks off into the distance, a refined show of emotion that speaks volumes.

I want to write this book honestly, and the truth is that when I was younger, I romanticized the lives of young people like Paul. Where I had comfort and security, I wanted freedom, danger, adventure, and the captivity of a London childhood couldn't provide these things. It's the case that many of us who live well enough to have the luxury of becoming sick are faced only with purely social, psychological challenges.

In January 2020, whilst filming *ANIMAL*, we visited a rural village in Marsabit, Kenya. Early one morning, we helped some of the women collect water from a well outside the village. To them, nature is new every morning. At home I can fumble my way to the tap with my eyes closed, knowing that I won't encounter an elephant between the fridge and the sink, but that day every step threw up clouds of red dust, dried by the beating sun, and you never knew what you might encounter on the path ahead. The village children watched on, giggling with bemused perplexity, as our weak Western shoulders buckled under the weight of the water. The cans frequently slipped to the ground, and in the end, the women tipped our water out and we carried near-empty containers. It was funny at the time because no survival value was being placed on our efforts, but if I'd had a young family waiting for me at home and I couldn't bring them water, that'd be it.

The reason I am turning first to the hardships and horror of water security is not to make you feel hopeless, or because I think you're not already aware. In fact, you probably knew of all this from a very young age. The old tropes are familiar and overused, misrepresenting the African continent and its inhabitants as a homogeneous and hapless mass who need a 'white saviour'. I am telling you because we are all part of the problem, and so we are all part of the solution. The tragedy is not only that people must collect water in such a difficult manner, but that those who don't have to (us) are making it harder for those who do.

Kenyan women and children spend up to one third of their day fetching water in the hot sun. But the walks are becoming longer and more dangerous as the water becomes scarcer. 'Where you used to dig thirty-foot wells, there would always be water; now you dig a hundred and sixty feet, and still . . . no water,' Paul tells me. Sometimes the water delivery is turned off completely.

This morning, I visited a shop near my house. As I walked down the aisles, I passed stacks of Volvic water, Vita Coco water, watermelon-flavoured water, electrolyte-enhanced water, 'eliminate cellulite' birch water, reverse osmosis deionized water. For an element that falls out of the sky and is necessary for life, we have done very well rebranding it as a commodity. It was hard not to feel cynical about this when I spoke to Paul, and he described how his mum would walk for miles to fill a plastic pitcher, then trudge home, only to repeat the exercise again and again. Although he doesn't go with her, other children make the arduous walk.

Cheru is five. Each morning, after she has finished her warm-milk tea, she ties a kettle onto her back with frayed rope and sets off with her older sister Dina. The round trip can take three and a half hours, and every extra moment they are out in the sun, it climbs higher and burns brighter in the vast expanse of Kenyan sky. She's small, much smaller than many of the others, and has to stop often to roll her shoulders and linger in the shade before forcing herself to stumble onwards. When they arrive, she scoops the muddied water into the rickety aluminium kettle, which holds only enough water for her morning tea – if there even is water that day, that is.

Paul tells me that 'for centuries, from traditional to modern set-up, women have acted as the primary and natural resource managers by procuring water, firewood, managing waste and providing health care'. In fact, at least half of the world's food is grown by female farmers (and this rises to 80 per cent in some African countries). Although in developed countries the link is obscured by easy access to water for almost everyone, in developing countries where gender norms are more rigidly enforced, the lives of many women depend on the fluctuating seasons and the height of the water table.

For example, as Paul tells me, 'In the traditional African set-up, women are the main custodians of the environment. They are the cooks, gardeners and healers. They pass the knowledge of the mentioned roles from one daughter to another; and her daughter to her granddaughter. Even though in some communities they have less social power than men, they have influence on the usage and management of natural resources' – which I would say is the greatest power of all.

However, here is the other side of the coin. Women control the water, much of the agriculture and the energy, but as they do so, many men try to control them.

It was dusk on 14 April 2014. At a secondary school in the sleepy town of Chibok, Nigeria, hundreds of schoolgirls were staying overnight to prepare for exams. The sleeping silence in the dormitories was shattered by the shouts of militants from the terrorist group Boko Haram (which translates roughly to 'Western education is sinful'). The militants opposed the Western form of education the girls were receiving, so as an act of resistance and hostility, that night they kidnapped more than 250 girls. It was a campaign of terror, and it worked. Some of the girls rolled from the trucks and ran into the African bush. Others were later released. To this day, though, many have yet to be found, and their locations remain undisclosed. Some of the girls chronicled the horrors in a succession of diary entries. One entry reads, 'The militants handed me a blade and issued a chilling ultimatum: "Cut off the girls' heads, or lose your own." We are begging them. We are crying. They said if we ran away, they were going to cut off our necks.'

To hide the diaries from their captors, the girls would bury the notebooks in the ground until nightfall, or carry them in their underwear. They are more than personal recollections; they are heart-wrenching testimonies to the violence that girls and women face at the hands of men.

At the same time this was happening, on the other side of Nigeria twenty-year-old Oladosu Adenike was studying. She is now a climate activist and self-proclaimed 'eco-feminist'. We speak over Zoom. I am in London, Oladosu in Nigeria. The connection is bad and her words are fragmented and fuzzy, but I hang on to every single one; the story is chilling, and very personal for Oladosu.

She explains to me how the climate crisis allowed for the growth of Boko Haram. Lake Chad was once a thriving source of food and economic opportunity for thirty million people. At dusk, townspeople would swim and wash in the blue-green water as fishermen cast their thick nets into the lake's depths. Today, drought and global warming mean it has shrunk by over 90 per cent, from a swollen body to a shrivelled scar, leading to struggle and insecurity. When humans find themselves insecure and struggling, they will do what needs to be done to find purpose and stability. There is a thin line between survival and evil, and Boko Haram have made many stumble across that line.

Oladosu speaks with passion and anger. 'It is psychological for me,' she says. 'It's hitting us women hard. The climate crisis takes the face of a woman and so we cannot solve it if the welfare of women and girls is left out. Gender-based violence cannot be solved if the various environmental situations are not solved. I have a dream, a dream where every girl and woman is equal to their brother, father and uncle in all ramifications; I strongly believe this can be achieved through eco-feminism.'

I had heard of stories similar to Oladosu's before. I wanted to know how an eco-feminist was nurtured. How did she channel the vulnerability and fear she had felt into such strength and defiance?

Oladosu grew up in Ogbomosho in south-western Nigeria, surrounded by cashew, mango and tobacco farms. If you dive deep enough, there is a tragic irony in the history of Ogbomosho. An early missionary described it as 'a walled city, the gates of which were closely watched by day and securely closed by night. It was an isolated, yet picturesque and well-watered city.' The inhabitants felt secure because of the walls that towered over them, but now the danger comes from every angle, be it from the chemical composition of the atmosphere, the aridness of the soil, or the burning sun fissuring the land. The danger cannot be confined; it is inescapable, and no walls will keep this village 'well-watered' – quite the opposite.

However, Oladosu speaks with tenderness of her childhood because it was an 'easier life'. Although 'population was increasing, and land was diminishing', she wasn't overly concerned. I think, in a way, many of the young people I've spoken to miss this oblivion. They wouldn't want to slip back into it, but they miss it. Childhood is loosely defined, but from everyone I've spoken to, it seems that the day they began to fret over the future was the day they began to leave their childhood behind. For Oladosu, this was two-tiered. Firstly it was a gradual realization of 'displaced peoples, farmer–herdsmen clashes, insecurity – all driven by climate change', combined with 'the increase in food price, floods sweeping away farmers' land, droughts affecting the yield of crops, and excessive rainfall'. Finally, it was the crashing realization about the link between Boko Haram and the environmental crisis.

I myself have learnt about the tragedies of the environmental crisis through a rectangular bluish screen. I have said in speeches that my hypothetical children's lives are being compromised and my generation are being handed a 'poisoned chalice'. Paul and Oladosu don't have the privilege of getting to worry about the world their hypothetical children may inherit; they are already living at the heart of the crisis. Large, looming and ever-growing before their eyes. However, a common thread unites us. We are a part of the same movement, the same proverbial fight.

This fight, morally and scientifically, should fall disproportionately upon adults. Psychoanalyst Erik Erikson defined eight major stages of development in life: 40–65 years of age is the stage of 'generativity', at which people generally develop a sense of being a part of the bigger picture. It is when people begin to make their mark on the world by creating or nurturing things that will outlast them.

However – and I say this as a teenager, not a scientist – when you're young, you believe you can overcome the obstacles you face in accordance with your desires. Maybe this is why the youth movement has snowballed. Generally, when you're young, you believe you must change the world to fit in with your vision of how it should be. As you grow up, you believe you must change your vision to fit with how the world is. It's this elusive conviction in how the world *should* be that has driven Paul and Oladosu and millions of other young people towards action.

Another of these young people is Autumn Peltier. Her story begins 9,000 miles away from Paul and Oladosu, on a great expansive body of fresh water called Lake Huron, one of the five Great Lakes of North America. She was born in 2004 and her activism is rooted in ancestral rivers of wisdom, by which I am engulfed as I plunge deeper into her story. It begins with Josephine Mandamin, Autumn's great-aunt. In most portraits, she appears warm and smiling, but she has that intense, defiant gaze of someone who holds the world together. And, really, that's what she did. As an elder and grandmother, with a copper pail of water in one hand and a staff in the other, she walked round each of the Great Lakes, the equivalent distance of half the earth's circumference, to raise awareness of the need to protect, conserve and respect one of our planet's most precious resources: water.

Benton Banai, another activist, summarized indigenous beliefs about the earth and women's connection to water when he said, 'She is called Mother Earth because from Her came all living things. Water is Her life blood. It flows through Her, nourishes Her, and purifies Her.' You might be uncomfortable right now. That's okay. Many of us have been raised in a culture that creates rigid boundaries between humankind and nature. Giving the earth personhood makes us squirm, makes us question the sanity of those doing so. But I ask you, as you hold these pages in your hand, to allow those borders to dissolve. It's going to be necessary as we go on to do so. As child-bearers, indigenous women feel they have a sacred connection to the spirit of water, and so they hold the responsibility to protect and nurture it.

Autumn referenced this in a speech at the UN: 'When you ask the question about why water is so sacred, it's not just because we need it, and nothing can survive without water. It's because for years and years our ancestors have passed on traditional oral knowledge that our water is alive, and our water has a spirit. Our first water teaching comes from within our own mother. We literally live in water for nine months, floating in that sacred water that gives us life. We can't live in our mother's womb without water. As a foetus, we need that sacred water for development. The sacred significance is that my mother comes from her mother's water, my grandmother comes from her mother's water, and my great-great-grandmother comes from her mother's water.'

So why is Autumn speaking at the UN and Josephine walking thousands of miles for water?

Before the Americas were colonized, indigenous women had much more political authority within their communities. As settler colonists moved in, much of the high regard for the beliefs and passions of indigenous women was torn to shreds. Autumn and Josephine's activism is partly reclaiming the positive aspects of indigenous women's traditions, such as the respect for and protection of water.

The campaign to protect water is akin to feminism in

Josephine's eyes. In an interview she describes the destruction of earth: 'She is being polluted, she is being prostituted, she is being sold. All that is happening to her is happening to us women now. So, when I think about how we as women have to pick up our bundles, we have to really think about how important it is that we really know who we are as women. That we are very powerful women, we can be very instrumental in how things are changing.'

She's right, things are changing. When Josephine was a girl, time was suspended as her days were spent plunging into the clear lake, rejoicing in its unsuspected depths, its crystalline shallows, singing in the summer sun and leaving the weight of the land behind. 'When you first look at the lakes and see the majesty of the water, you want it to stay like that for ever and ever and ever for the generations to come.' But just as the clarity and purity of childhood is muddied by age, the lakes too have been subjected to the same process at the hands of humans.

I want to introduce you to these lakes. To Josephine, they are like family. There are five of them, Superior, Huron, Michigan, Erie and Ontario, and they have a total surface area of 94,600 square miles (containing over 20 per cent of the world's fresh water), making them the largest fresh-water system on the planet. Each lake has its own distinctive qualities, and Josephine knows them intimately. 'Lake Superior is kind and gentle, but she can also be treacherous and unforgiving. Lake Ontario, she's special to me. That's where I was born . . . but she was the hardest to walk around.' It wasn't the terrain or the weather, it was the fact that as she walked, she had to push through stench clouds from rotting fish that had died in a soup of pollution. The once-sublime lake she so loved was now smeared in thick green slime and algae. Once, she would have peeled off her socks and cooled her blistered feet in the lakes, but these days she was afraid to do that. The water was full of chemicals and toxins. 'It was metallic and poisonous,' she recalls.

The Canadian Great Lakes suffer from four main sources of pollution: manufacturing, agriculture, heavy industry and plastic. Climate change is also causing them to warm, which reduces the surface levels and affects the distribution of aquatic life. Josephine's activism is more than just a defiance of local injustice; she's taking a stand against global injustice that extends far beyond the Great Lakes.

So far, we've journeyed to different continents and met unique women affected in various ways by the environmental crisis, not to see them through a lens of pity and victimhood, but to lead me to this point. When asked what you think the most effective ways to combat climate change are, your answer might be something along the lines of installing solar panels on rooftops, building wind turbines, going vegan or driving electric cars. In 2017, Project Drawdown, a coalition of researchers and scientists, came together to identify the most effective way to not only halt emissions, but somehow cause a decline in the amount of greenhouse gas in the atmosphere. The results showed that, ranked according to efficacy, educating girls was above most of the solutions usually proposed, including rooftop solar panels and electric vehicles.

I've asked myself why there is such a thick and almost unbroken silence around the idea that educating girls could be the most moral, efficient and impactful way to fight the climate crisis. Why aren't media outlets all over the world speaking about this? Probably because the modern concept of sustainability is led by Western nations, where it is taken for granted that girls' education is available for all. Around the world, however 130 million girls do *not* have access to education. As we work to reimagine transportation and reduce emissions, we should also be committing time and energy to fight this fundamental battle.

I want you to meet a girl who stood on the brink of her education being ripped away from her. Let's travel to Rajasthan, northern India, where Payal Jangid was born. It is India's largest state, peppered with opulent palaces and hill forts. It is also a place with villages where most girls are married before the age of fifteen. Imagine all the untold stories, snatched possibilities and lives left unlived when a young girl is bound into a marriage and belongs to someone else before she has even had time to know herself.

Child marriage is a factor contributing to environmental destruction, but it is also caused by it. In Payal's village, and all around the world, as the temperatures rise, floods drown crops and the seasons become more unpredictable, families find it increasingly difficult to feed and educate their children. As a solution, they sacrifice a girl or two to marriage so there are fewer mouths to feed. In countries like South Sudan, many people are using their younger daughters as a means to acquire cattle, out of hunger and desperation because their crops are failing due to climate change.

Payal was expected to marry at thirteen. She was in school, two years before the dreaded date arrived, when a group of activists came to speak about the importance of education for girls. For Payal, puberty and marriage were already both bounding over the horizon like a stampede of unstoppable racehorses. She was just a child but would soon become a vessel ripped from education and impregnated. Her parents had already sat her down and spelled out the facts for her. However, when she heard the activists' talk, it awakened some kind of resistance, a fury within. She begged her parents to change their minds. After listening to her arguments, they agreed, but she didn't stop there. She made sure her fifteen-year-old sister wasn't channelled into marriage either, before going around to every door in the village handing out flyers and staging protests to protect her friends from the same fate.

It worked. Her village, Hinsla, transformed from a hotspot of child marriage to being completely free of it. Rather than flitting from being her parents' child to her husband's property as expected, Payal became a tireless crusader against forced marriage, and for girls' education too. She became president of her village's children's parliament, and together the village youth raised their voices against the injustices they faced. The girls stayed in education, the adults listened to the children, and the neighbouring villages began to follow Hinsla's lead.

What would happen if every village copied Hinsla? As well as being a cause in its own right, the links between educating all children, especially girls, and protecting the environment are clear:

1. If we achieved secondary education for girls across the globe tomorrow, by 2050 that would prevent a population increase of 1.5 billion more people than if access to education remains as it is today (that's equivalent to more than 85 gigatonnes of carbon emissions avoided).
2. Educating girls naturally leads to more women in leadership positions, which studies have shown gives rise to businesses and countries with greener credentials. Countries with more women in parliament are more likely to create protected land and introduce stricter policies around protecting the environment, whilst also ratifying more international environmental treaties. Plus it has been shown that countries where women have a higher social and political status have lower climate footprints and produce fewer carbon emissions.
3. Education provides girls with the skills needed to ensure a just transition to a much more resilient climate economy. Currently, women sit in less than 20 per cent of occupations in the

clean energy sector. By investing in girls' STEM education (which is where green industries are beginning to emerge), we will ensure that women not only participate but actually help lead innovation into green and climate-resilient technologies.

Manifesto for change

- Put pressure on the government. Around the world, governments can help to save fresh water and prevent water pollution by:
    - measuring water use and setting targets to reduce it
    - obligating large companies to measure and manage the amount of water resources they use
    - encouraging lower-water diets (including reduced meat consumption)
    - supporting industry to make water-intensive products last longer
    - providing consumers with the tools to understand the water impacts of the things they buy
    - making laws to increase water recycling
    - preventing people and companies from polluting waterways by making laws against toxic chemicals that could pollute our soils.*

* Suggestions from Friends of the Earth.

For the individual:

- Eat in season and beware of what you eat. Harvesting crops like avocados year-round and rearing animals for meat and dairy is incredibly water-intensive. It requires roughly 15,500 litres of water to produce 1 kilogram of beef, and meat production in total demands almost 2,300 trillion tonnes of water every year (about 72 million litres a second). By cutting down on meat and dairy and eating seasonal vegetables, you'll be helping to conserve water, whilst also reducing your carbon footprint at the same time.
- Use your dishwasher and fill it up before running it. Dishwashers are four times more efficient than washing up by hand, so this way you'll use less water.
- Take a shower instead of a bath. A five-minute shower uses about 40 litres of water, which is approximately half the volume of a standard bath.
- Don't fund 'water-grabbers'. Land grabbing is a process where much of the world's agricultural land is being turned from small-scale farming to large-scale commercial farms. Huge tracts of land are purchased by organizations, usually in countries where the organization isn't even located. These large organizations and companies sometimes deny local people access to water, pollute the local watercourses and

exhaust supplies. To help stop this, make sure you know where your savings or pension are invested. It's absurd that money put aside for our future should be used to make that very future impossible. You can also check how the companies you use treat local water sources – avoid them if they contribute towards water wastage and contamination.

Chapter 7

Rewilding the World, Rewilding Ourselves

Rediscovering the wonders and wilds of nature

We were waiting for the wolves.

A heavy, bruised-plum dusk sky. Beams of moonlight spilling over the hills and pooling in the field where we sat. The leaves spinning and crackling in the wind felt as loud as thunder as we strained our ears, listening. We were in Jura, France, filming for *ANIMAL*, about to meet a shepherd who was adjusting to these new predators slinking across the land. We had spent the day tracking the wolves, looking for their prints and watching scientists marvel at clusters of poo, so we knew they were there. At times, as we sat amidst the steaming flock, we could hear haunting howls cutting across the valley, but the wolves themselves seemed to have melted into the landscape entirely, leaving not even a shadow behind.

Perhaps that's why they've been so successful. There are roughly 530 wolves in France now, mostly living in the Alps and the south-east of the country. In the ninth century, France created the 'royal office of the Luparii' to rid the country of the predators. It was only in the twentieth century, with improved technology such as rifles and the use of new poisons, that the job was said to be complete and the very last wolf disappeared from the mainland. But then, as humans often do, we forgot about the 'problems' of the past.

In the pale morning light of modernity, people are

being swept into cities, land use patterns are changing on a grand scale and much of European farmland is tumbling into disuse. Land once farmed now goes unclaimed, unploughed and, therefore, unrestrained. Some estimates predict that by 2030, an area the size of Italy will have been abandoned within the EU alone. When we push wilderness out of new areas, especially our cities, bursts of nature bubble up elsewhere. Thorny crops and shrubs are hoisted from the earth, tiny trees take root. Insect life bristles among vegetation and sends humming vibrations across the land. Songbirds sing up and mammals nestle in, vegetation claims tumbledown gates, abandoned barns and houses. Then the megafauna come. Deer strut across shabby, ragged pastures. Wild boar rootle in the undergrowth, gleefully inhaling the absence of human influence upon the land. Finally, slinking through the shadows, belly to the ground, largely unnoticed at first, come the carnivores. Lynx, brown bears, wolves.

In France, they crossed the Alps from Italy, sending ripples of anger through the rural communities, especially the farmers. Wolves. That word alone is enough to awaken the most primal fear within us. We've been besieged by stories since we were children. Taught to fear this species. I remember my cousin, when he was three, thinking of himself as the boldest, bravest toddler, and yet he would come slinking down the stairs in the dark, whimpering and shivering at the 'big bad wolf' that had just hounded him in his dreams. So when a fear of wolves is so deeply embedded in our cultural attitude, how can we learn to protect and love them? Or at least learn to peacefully coexist with them?

At this point, I'd like you to stop and decide your stance on this issue. Think about your arguments, your reasoning and then bundle them into a corner and approach this topic free from what culture has had you believe. When I went out to Jura, I went with an ecologist's mind, adamant that rewilding (restoring natural ecosystems to the point where processes that have been halted by humans can reoccur, higher levels of biodiversity can return, and natural change is allowed to happen) was the best thing possible. However, there I had to listen to the farmers' stories, and in order to *truly listen* to someone's story, you have to approach disentangled from the ideals of yourself and others.

The first step is to recognize that the idea of rewilding proposed by environmentalists is not just a romantic proposition, it's a necessary one. Romanticizing it can be tempting and appealing, seeing the thrill of wildness and possibility that wolves would stitch into the land, but there is much more to it than that. There is an ecological benefit, or even necessity, in some landscapes that large megafauna, especially carnivores like wolves, provide.

Deer don't seem capable of the destruction they cause. They are regal figures, red deer, with their antlers coiffed in russet and gold silk and their balletic tiptoeing gait. I've spent many hours roaming Richmond Park, a bubble of wild in London, with the moon beaming down on me as I listen to the roars of the stags during rutting season and watch their ghostly figures slip through the ferns. However, during certain seasons, there is a cull. 'Males in February, females in autumn', they say. Then the meat goes to game wholesalers. Although I feel incredibly sentimental about the deer, they have to be managed or they would overrun the park, destroying the ecosystem and competing for resources until they bit every blade of grass down to the quick and died of starvation anyway. In an enclosed park, rangers can resort to sterilization programmes, but in the sweeping wilds of the Highlands of Scotland, that's impossible. Due to the lack of natural predators, the red deer population has more than doubled since 1959, rising from 155,000 to an estimated 400,000.

Some people believe that reintroducing predators like wolves is the solution. One such example is 'the Wolf Man', aka Paul Lister, who is based in the vast expanse of the Alladale Wilderness Reserve in Scotland and wants to establish two packs of ten wolves on his land in the next few years, followed by bear and lynx soon after.

The reserve shivers with life, unlike the glacial expanses of ecological desert that dominate much of the UK, which we are conditioned to believe are wild. The riverbed is swollen with birch and aspen, bilberries bloom and Scots pines straddle the hillsides. Oh, and the wildlife – the wildlife! Otters play, raptors and corvids wheel and dive across the wide sky. There are squirrels and pipistrelle bats, badgers and pine martens, salmon and trout . . . and water voles!

Lister believes that bringing in wolves would help the local economy by attracting roughly 20,000 visitors a year. As George Monbiot discusses in his book *Feral*, the reintroduction of sea eagles on the Isle of Mull has done something similar, bringing £5 million a year to the island

and supporting more than 110 jobs. In Sussex, the Knepp Castle estate transformed its unprofitable wheat farming into rewilded, largely self-wilded land, and now hosts some of Britain's highest populations of nightingales, purple emperor butterflies and turtle doves. It has become more profitable than when it was functioning as a farm, through ecotourism, accommodation and selling high-grade meat. In other parts of Europe where bison and wolves are being reintroduced, shepherds are learning to live along-side them through modern technology, as well as traditional techniques such as keeping large dogs to protect their flocks, and using collars that monitor the sheep's heart rate and contact the farmer if they show fear.

One of the greatest rewilding and conservation success stories began twenty-five years ago in Yellowstone National Park in the United States. The story is told in a YouTube video, 'How Wolves Change Rivers', which now has over forty-five million views as people are captivated by the power one simple ecosystem change can have. The grey wolf, which had once roamed from the Arctic to Mexico, had been completely eradicated by the park's employees as part of the policy to rid it of all predators. Then, in 1995, the number of deer was so high, and they had grazed the vegetation down so severely, that the rangers decided to reintroduce wolves.

The wolves reduced the number of deer, but much more significantly, the deer's behaviour changed radically. They began to avoid the gorges and valleys where they could be caught more easily. Without heavy grazing, bare valley sides quickly became dense forests. Then the birds started to move in. The number of beavers increased, as they like to eat the trees. The dams built by the beavers provided habitats for fish, reptiles, amphibians, otters and muskrats. The wolves also killed coyotes, so with fewer predators, the mice and rabbit populations boomed. This allowed for more hawks, badgers, foxes and weasels. The carrion left by the wolves increased bear populations. The bears then reduced the deer populations, thus reinforcing this positive cycle. More strangely though, the behaviour of the rivers also changed. The regenerating forests stabilized the banks, so there was less soil erosion, meaning they collapsed less often and meandered less. This long chain of events, triggered by the addition or removal of top predators, is called a trophic cascade. It proves how important the web of interactions is, and how much damage we do by removing species and messing up the natural balance.

But was the occurrence at Yellowstone just an anomaly? Is rewilding always the solution? To dive deeper, I spoke with Holly Gillibrand, a Scottish force of nature who had become my friend after we'd worked together on several projects.

Holly was just nine when she arrived in Scotland. She'd moved there from Tasmania, a land of serried mountains, overhung by clouds, with heavy seas thrashing against the coast. Half of the state is classified as a World Heritage area or national park. She remembers that when splashing up the streams with her neighbour's dog, wombats would wobble up to her, so docile that she could run her fingers through their fur like a comb. 'It wasn't something I thought or worried about much,' she tells me.

'The wilderness was just a part of me, and taking that part away was unimaginable.'

When you leave a landscape, it doesn't necessarily leave you, so Holly found herself comparing the forested hills of Tasmania with the bare bones of Scotland, its broken trees like wounded soldiers in the stripped-naked land. Although she lives among the hills, she explains that she hates hill walking. It's what she doesn't see there that bothers her. Many times she has walked for hour upon hour, over and through the landscape, and seen only a scattering of deer, nothing else.

'Why? Scotland is supposed to be the last great wilderness of the UK, isn't it?' I ask.

She tells me about the grouse moors; huge swathes of land trimmed to the quick and managed like golf courses to provide a stomping ground for shoots. They're estimated to cover somewhere between 1 and 1.5 million hectares, which means that almost one fifth of Scotland is managed for grouse shooting. 'People often talk about conservation, but we can't just conserve dwindling species and degraded habitats, we actually need to restore the abundance of nature and revive ecosystem processes through restoration . . . and rewilding.'

Holly is well integrated into the rewilding community in Scotland. She's the 'biodiversity voice' for Heal Rewilding, announcing their scientific policies, and she presented to the world a rare breeding pair of reintroduced ospreys live on Chris Packham's channel. So I pose the big question to her: 'Do you think we're ready to begin reintroducing carnivores?'

She pauses for an instant. 'Not completely,' she says. I didn't know exactly what answer I was expecting, but that wasn't it. 'The thing is,' she explains, 'we could do it theoretically, but I just don't think everyone is ready for it yet. For many of those whose income comes from the land, there is still that ingrained Victorian perspective of shooting everything that moves. If we can't even coexist yet with smaller species like beavers, where will we stand with carnivores and larger predators like wolves?'

Our physical landscape is ready. In some places, like Scotland, reintroducing predators would require fencing off large areas at first, but we could do it. However, it's the mental landscape of the people that remains unready and hostile to the idea of change. Our imaginations have wiped the hills of history and swallowed the potential for possibility so that we remain adamant that we could not live in a wilder world. As Holly says, we're stuck in a Victorian mindset, viewing nature as something to be brought inside and arranged nicely in a vase, supple and tame under human fingers.

So how do we get people to fall in love with the wild as it is now, but also with what it could be? This would involve us loving the laced diamonds on an adder's back, the sharp dive of an eagle, and the sweet chatter of a red squirrel. Only then can we imagine a future where rare bubbles of joy like those found in nature can be experienced continuously. That way, nature would no longer be confined to fragmented pockets but allowed to flourish in, around and between us. Springing buoyantly up on verges and in parks, on roofs, in front gardens and on street corners.

I've discussed the ecological benefit of rewilding with megafauna and vegetation, but what about the benefit of rewilding humans? In Chapter 5 I spoke about nature deficit disorder afflicting young people, and as you're reading this, I am assuming you're already in a place where you respect and admire the natural world. However, the natural world is imbued with magic and possibilities that even many self-proclaimed naturalists fail to see. Many who say they enjoy being outdoors may actually mean that they like the orderly line of carrot tops in their vegetable plot, the soft whisper of a butterfly's wings, or the hug felt from the scent of a sweet rose as delectable as honey – but do they love *self-wilded* nature? Not the beauty of a micromanaged flower bed, but instead, the brilliant, messy wild with all its rugged imperfections? Let your grass grow tall and seed, leave the leaves, forget the chemicals, build a brush pile. If we all embrace a new paradigm of relaxed gardening for the ecology not just the aesthetic, and realize that one person's mess is another person's gift to wildlife, we could be living in pockets of abundance and have no need to travel far and wide to catch a glimpse of wildlife.

In fact, a wilder future can be coaxed back to life, and kindled even, on just a small strip of derelict land.

Staffordshire is better known for bull terriers and pottery than for being a hub of environmentalism, but two seventeen-year-olds are quietly creating there what many of us are only daring to imagine. A modern Jurassic Park is emerging amidst a patchwork of cabins, greenhouses and a pond. If you peer into the home-made enclosures, you'll see emerald greens and the beady black eyes of reptiles long gone from England but that perhaps one day soon will be slithering and hopping across those 'green and pleasant' lands. It all began when young Harvey Tweats started breeding stick insects in his room. He also collected frogspawn, and reared butterfly and moth caterpillars.

I spoke to Harvey on a school day, when both of us were learning from home, and squeezed our call in between lessons. He was in a box room coloured the teal of a duck's egg, and had Swiss cheese plants stretching up the wall like beanstalks. Due to the nature of what he has to do (cleaning cages and focusing on the hands-on side to conservation), I was expecting someone very technical and to-the-point. He was exactly that, except he would gracefully sweep from discussing a snake in a cage outside to the crisis of imagination our society faces, preventing us from building a different future.

He explained how, as a teenager, he met Derek Gow, who invited him on work experience studying the impact of beavers on a river valley near his home. Beavers fell into extinction in Britain somewhere in or around the seventeenth century, taking other species of insects and amphibians with them. Derek Gow is called 'the Beaver Man' in conservation circles. Apparently, with his large frame and round, bearded face, he bears an uncanny resemblance to the creatures, which he'd probably take as a compliment, considering his reverence for them. In the wooded ravine on his land, he has released dozens of beavers into wild expansive enclosures and allowed them to manipulate the environment in a way they should always have been doing, tumbling through the silvery streams and transforming the landscape into a cascade of life. It soon went from having no amphibians at all to thousands upon thousands.

Harvey recalls standing on a beaver dam and realizing how the beavers were ecosystem engineers. The webbed labyrinths of their mud-caked dams had created marshy wetlands for amphibians, insects, birds, mammals and reptiles.

With Tom, a lifelong friend, he acquired a licence and began to breed reptiles on a strip of land near their homes, with the aim of rewilding them. Both Harvey and Tom, I should add, are doing their A levels at the moment. We spoke about school briefly, and Harvey told me how it was a nightmare to be confined to the dry lines of ink stamped into pages of a textbook or the screen of a whiteboard studying the Krebs cycle or nitrogen-fixing bacteria, when the living, crawling, beating biology was waiting in his back garden, calling him away.

I ask him why he's focusing on the smaller species. Surely reintroducing wolves and beavers should be the main focus to change things as quickly and profoundly as possible?

'There was a study conducted in America that took ten hectares of healthy biodiverse marshland,' he tells me. 'They counted and weighed every single amphibian and the total weight was the same as a black rhino. So although many people see amphibians as small, together they assemble into this massive superorganism that controls disease, feeds so many species, and allows the ecosystem to function. So when people approach me with this arbitrary view that a frog is small and therefore has a lesser impact, I tell them that each different species inhabits slightly different niches in the ecosystem and so they open up availability of food and existence for other species. People are reintroducing white storks in Knepp, Sussex, but before that they have to think about returning food to the landscape . . . the amphibians and reptiles.'

And beyond their ecological importance, imagine the fascination that would grow out of having these species back on the land. The children who would fall in love with nature again, searching in the backwaters for elusive turtles and living in the continual curiosity of wildlife, never knowing what may be around the corner, in a tree, or in the murky depths of your pond.

The reason for reintroducing functionally extinct species, and not simply conserving what we have left, as Holly said, is because we have lost so much. In less than a human lifetime, more than two thirds of the vertebrate wildlife of the world has been wiped out, and a third of the world's natural resources are degraded today, many severely damaged beyond unassisted recovery.

One of the powerful ways to encourage people to move towards this future and to cultivate deep respect and love for the natural world is to help them envision it, because imagination is the precursor to creation. As I can't transport you to a rewilded future, I want to give you a lens to peer through instead, Dara McAnulty's lens, because no one will fail to fall in love with the *wild* wild if they see it as he does.

As a five-year-old, overwhelmed by the yelling cacophony of piercing noise that was Belfast, Dara was told that he had Asperger's, which, rather than being a limiting factor for him, was like a magnifying glass that intensified his experiences out in nature. However, he recalls how in Belfast, 'Everything about the constant buzzing was a struggle for me, I couldn't fade out the noise to hear what mattered. The birdsong was sullied by aeroplanes, the relentless traffic; you couldn't escape it, even when you tried. My parents did try but we were all feeling the pressing down of a busy life. My two siblings are also autistic, as is my mum. Something had to change.' They decided to move out to Fermanagh, to live alongside the cloaked conifers of Big Dog Forest, and like a tree, Dara became firmly rooted in the landscape around him. He described how 'the valley sings, heaves and rests', caterpillars move 'like slow-motion accordions', whilst a goshawk chick looks 'like an autumn forest rolled in the first snows of winter'.

Yet sometimes when you focus on one part of your life and it begins to swell with the best kind of happiness, other bits begin to crumble. At school, Dara was bullied, mercilessly. Children can be harsh and scathing to those who are enamoured with something different, something outside the box of convention.

When you're young, you are bombarded by media depictions of 'falling in love', and it is always in the most anthropocentric meaning of the term. I wouldn't hesitate to say that Dara fell in love with the landscape. At a time when he could not imagine acceptance from most of the humans around him, the great green countryside opened its arms for him to tumble into, and he did. He was the nature boy, fascinated by shreds of dappled light and the blue of a goshawk's eyes. He was the boy whose heart fluttered at the sight of a bird and who could crouch on his haunches watching a stream of ants for hours, yet couldn't stand the noise and clutter of classrooms.

Speaking to me on our call, he laughs easily, talks about the world with an almost unmatched passion, and I wish I could bottle some of the nature love he is exuding and send it around the globe. I wondered how he changed from the bullied boy who was once told that he would never be able to string a paragraph together to the award-winning sixteen-year-old author he is today. He explained that he stopped trying to repress himself and his passions to please others, and instead embraced them. He started a satellite-tag project to conserve birds of prey, began an eco group at school, and took lead roles in organizing and speaking at the climate strikes.

In a sort of frenzied desperation to discover the essence of his passion and instil it in others, I asked him, 'Why do you think you don't really just observe nature, but rather, identify with it? How do we infuse it with wonder for others?' He doesn't miss a beat. 'We don't need to infuse it with wonder, we just need to pay attention, because the wonder already exists.' Everyone is born a naturalist, he tells me, with a curiosity to match the best scientists. Then the moment you hit puberty, you're pushed into dry, uninspiring books on history and algebra. Catching the 'nature bug' years later is not about discovering your love of nature, it's about rediscovering it.

Dara tells me that as a toddler, he was afraid whenever the thick ice would creep across the pond, because he thought it would freeze all the life underneath. One day he ventured through the long grass brandishing a stick, and with arm held high, he brought it crashing down on the ice. In that moment, he realized that the ice was only a thin layer on top of the pond, and life persisted in the watery world below. When he was a bit older, he taught himself about complex hydrogen bonds and the science behind the changes in state happening on the pond. And that's what we must all do.

When you see a large long-legged bird lifting with the grace of a feather dancing in the wind, ask yourself how. Rewilding yourself means taking notice of the small bubbles of life everywhere, and the continual quotidian wonders around you. Much of the rest of the world isn't as dull and drizzly as the UK can be, and yet even though we have ravaged and desecrated much of our wildlife, Dara shows how this 'bleak' land still teems with wonder. All it takes is a little bit of curiosity and a pinch of awe, and you too will soon see the world brimming with marvels.

Halfway through our call, Dara knocks some things off his desk in a flurry of activity. 'Is that . . .' He jumps up to peer through the window, and I stare at the empty chair. 'Is it . . . is that . . . A buzzard!' he shouts. I try to wrap up the call quickly, for it is almost the hour before twilight, and the sun is streaking the sky in brilliant watercolours.

It's Dara's favourite time of day, when, as he says, 'the night sky seems to chase the sun out of the sky and you can't really see the sun's light but you know it's just underneath the horizon. At that time, the energy in the air is so charged. That's when the bats begin to come out, and you feel such a rush.'

As you can see, being an environmentalist is not all about sacrifices and bemoaning the fate of our planet. In fact, that's very rarely the motivation. We humans are a strange, long-legged, thin-skinned, largely naked species with big eyes and big foreheads, who are clumsy at birth and never seem to figure things out completely. One thing that is remarkable about us, though, is that we hope, imagine and dream. We are hard-wired to correlate forward motion with reward (not just with avoiding future harm), and so creating positive expectation is far more powerful than anticipating punishment.

Believe it or not, many environmentalists I've met are afraid to see the world as Dara does, or at least to express it the way he does. Why? They say it's because it might make us seem flaky. We will be dismissed as naïve and overly sentimental. What we need, they insist, is just more hard data, numbers and longer lists of statistics to frighten people into action.

I would argue that this stance is mistaken. By shying away from the idea of our planet as the wonderfully alive, magical place it is, we are endorsing the very disconnection and apathy that enables its destruction in the first place. Perhaps we don't want to reintroduce predators and carnivores because we like to be the top predator, to be superior. This attitude is a loaded gun pointed at the beating heart of this planet. Although it may make us feel powerfully human, even god-like, to dominate our world, when you really understand what we're losing and that it may never come back, you realize the frightening ease with which this domination makes us less human than ever.

This self-imposed role as the complex creatures who must rule above and out-compete other forms of life, both physically and intellectually, means we tend to overcomplicate things. If an answer is too easy, we tell ourselves it must be the wrong one. So when we hear about the massive amounts of carbon swirling around our atmosphere, we assemble the best scientists, engineers and inventors and spend millions on building machines to remove it, forgetting that this very 'machine' already exists, costs very little, and builds itself. It is called . . . a tree.

Recent scientific studies estimate that worldwide planting programmes could remove two thirds of all the emissions in the atmosphere resulting from human activities. There are 1.7 billion hectares of treeless land worldwide – about 11 per cent of the land on earth – that could be converted into carbon-sucking, wildlife-booming, oxygen-pumping oases. If you've planted a tree, you'll know that for best results it's not as simple as tossing a seedling into some soil and watering it. You need to ensure that the species fits the landscape, that the area and the timing are right. However, doing this bit of research beforehand and then getting stuck in will pay dividends for you and the planet. To set you off on your own restoration journey, I want to take you to a rugged plot of red soil in Nairobi in 2011.

A seven-year-old boy, Lesein Mutunkei, kitted out in oversized soccer gear, crumbles the last handful of soil over the roots of a small tree. The rising sun and lifting shadows throw golden copper light into the air, and he rises, smouldering with pride, as the young tree wavers in the breeze. Over the next few years, he overhears snippets of information about deforestation and climate breakdown. A webbing of his tree-planting memory and this new information begins to take shape in his mind, until one day, as he thunders down a soccer field, an idea breaches the surface and a plan crystallizes. If he can score this goal, he will plant a tree.

That was how Trees for Goals was formed. Knowing that he was going to do something good if he scored a goal motivated Lesein to play harder, but soon it became less about soccer and more about the feeling of soft earth under his fingers, of satisfaction at knowing he was a part of something much bigger.

Lesein began his own tree-planting and restoration projects after calculating that enough rainforest to fill thirty football pitches is hacked down by humans every minute. 'How did it feel planting trees alone in the face of such a huge challenge?' I asked him.

He tells me about an allegory by activist Wangari Maathai that she first related in the movie *Dirt!*: 'A huge forest is being consumed by a fire. All the animals in the forest come out and they are transfixed as they watch the forest burning, and because they feel so overwhelmed and powerless, they stand there doing nothing. The hummingbird decides, "I'm going to do something about the fire!" So he flies to the nearest stream and collects a droplet of water and returns to put it on the fire. He goes up and down, up and down, up and down, as fast as he can. In the meantime, all the other animals, much bigger animals like the elephant with a trunk that could bring much more water, are saying to the hummingbird, "What do you think you can do? You are so little and that fire is too big. Your wings are too little and your beak is too small to bring much water." But as they continue to discourage him, he turns to them without wasting any time and tells them, "I am doing the best I can." And I will be that hummingbird,' says Lesein. 'I will do the best I can.'

It turns out that sometimes when others see an individual making change, revolution becomes contagious. Lesein's school started Trees for Goals in their football, rugby and basketball teams, and even his football club adopted his idea. Soon they had planted nearly a thousand trees around Nairobi. Then the Ministry of Environment and Forestry noticed Lesein's campaign on social media and invited him to meet their board and the minister. They discussed how his approach could engage young people with the environment and agreed to provide him with trees.

He was soon invited to the UN Youth Climate Summit in New York. He boarded the plane for his first trip outside Kenya, tracing his fingers over the aeroplane window, the snow-capped mountains and tumultuous seas, like he was running them over the topographical map in school.

At fifteen years old, he was one of the youngest at the conference and was determined to do even more after meeting other young people taking action. When he returned to Kenya, he was invited to meet the country's president and plant trees with him.

As we're ending our call, I ask Lesein about his hopes for the future. 'I hope we can get football teams all around the world, including FIFA, to adopt this approach. After all, the climate crisis is a universal crisis and football is a universal game.'

Although planting one tree is not going to reverse the mass destruction of forests, often it is through these smaller, personal steps that one enters the world of sustainability and environmentalism. All of these small actions begin to surge together and send out ripples that inspire others and have an impact on political and economic structures. Speaking to lots of activists has taught me that one of the main reasons people become burnt out and tired of activism is the fact that it requires you to always be railing against something, to always be fighting the system. Yes, we require transformational change in that large-scale, top-down way, but as influential climate activist Mary Heglar says, 'We can be overwhelmed by the complexity of the problem, or fall in love with the creativity of the solution' and the beauty of the future that hangs in the balance. How can you not fall in love with a wilder world beating with life?

Manifesto for change

Individual:

- If you have access to green spaces of your own, let them become wilder. Let your lawn grow, relax the weeding, and discuss with neighbours the possibility of opening up your garden boundaries to create corridors for wildlife. Remember that hedges are much more wildlife-friendly than fences. You can also create holes in the bottom to allow the wildlife through. Alternatively, join a community garden and help the space become wilder, or ask your local council to create a wild space in your local community.
- See your garden from a bird's perspective and plant at a mix of heights for different species. Long-tailed tits move above head height, whereas other species like wrens will dash in and out of dense cover closer to the ground. You can also put up nesting boxes in different positions on buildings and around the garden.
- Borrow/buy and read *Wilding: The Return of Nature to a British Farm* by Isabella Tree. Use her wealth of knowledge to start conversations with others about rewilding.

Large-scale:

- Encourage nature-friendly farming methods that work in harmony with the natural world – agroecology, circular agriculture, permaculture and organic farming – by buying from those that produce in this way.
- Support the growing number of organizations that are calling for 30 per cent of land and sea to be restored for nature by 2030: https://www.nationalgeographic.com/environment/2019/01/conservation-groups-call-for-protecting-30-percent-earth-2030/.
- Support the incentivization of the replanting of regionally appropriate mixed woodland or forest, for example through donating to organizations like World Land Trust.
- Campaign for governments to incentivize the rewilding of natural habitats. This can include wetlands, peatlands, grasslands, mangroves and areas of coastal sea grass.

Chapter 8

Environmentalism and Intersectionality

Shifting the narrative and making space at the table

Children all over the world are gripped by hacking coughs, wheezing through concrete lungs blackened by pollution. Most of these young people are, statistically, people of colour. In a 2018 study on air quality published in the *American Journal of Public Health*, researchers found that black residents 'had 1.54 higher burden than did the overall population' of exposure to particulate matter.

Here's one of the greatest problems: the stories being told about environmental activism and activists in the media are largely centred on conservationists, environmentalists and authors who speak from a privileged podium of wealth and whiteness. To understand the vicious and deadly cycle preventing us as a species from taking great leaps forward to protect our planet, we need to talk about how we silence the voices of those most affected.

Here's how it goes:

1. Societal racism, including public policies and general public attitudes, means people of colour are usually concentrated in specific areas, leading to residential segregation.
2. Neighbourhoods with more people of colour have, historically, experienced lower property values, and still do to this day.

3. As a result, land in those areas is cheaper for players in the industrial sector to acquire, meaning there are higher levels of mining, oil and gas extraction, leading to greater pollution.
4. Communities of colour are therefore usually concentrated in areas that face greater environmental threats, and find it harder to move to less environmentally harmful areas because of systemic political and social barriers. Some 68 per cent of black communities and nearly 40 per cent of Latinos live near a dangerously polluting power plant.*

Last week in school, the teacher showed us maps of American cities in 1919. They seemed innocuous at first. The suburbs were fragmented into blocks of colour. Sunflower yellow, bruised-plum purple and fiery red. The blocks of red were sparse and separate, always bundled into the corner of the maps. They were BIPOC (black, indigenous and other people of colour) communities. The term 'redlining' refers to 'the practice of denying a creditworthy applicant a loan for housing in a certain neighbourhood even though the applicant may otherwise be eligible for the loan'. This practice was outlawed in

* As a side note, in 2017, the American Petroleum Institute wrongly insisted that African Americans have more health issues, such as higher rates of asthma, because of genetics rather than pollution. There's no research on which to base these outlandish claims; it's just an example of the lengths industries will go to to absolve themselves of blame and guilt.

1968. Despite these families now having the option to move, they don't suddenly have the means or privilege to be able to afford the suburban homes that many white families purchase in the 'better' areas. Historically, many families have had to live in redlined areas for generations, worsening systemic disparities: many previously redlined areas are now near freeways, which exposes these individuals to higher levels of exhaust fumes and poorer air quality.

Genuine, nuanced and collaborative responses to inequality worldwide are rare. A surface-level 'collaborative' (and highly performative) response to inequality that the world saw recently was when individuals plastered their social media feeds with black squares in support of Black Lives Matter. This lazy display of allyship extinguished any illusion I had that people were willing to put in the work and help to create real structural change.

One particular voice I want to talk about is Choked Up, a campaigning group consisting of – as they describe themselves – brown and black teenagers who want the right to breathe clean air to be made law. Anjali Raman-Middleton, the sixteen-year-old founder of the movement, lives near the South Circular Road in London, where the blue summer sky seems congested and the groans and wails of vehicles are so loud that silence is almost a stranger. In 2013, her school friend Ella Kissi-Debrah died from an asthma attack on a day when pollution levels on that road were fatally high. In a landmark ruling, Ella has now become the first person in the UK to have air pollution officially recognized by a coroner as a cause of death. I discussed Ella's story in detail in Chapter 2, but it is Anjali and her friends who now bear the weight of that legacy, pushing desperately for change.

In Anjali's words: 'Behind the scenes there has been a concerted campaign from vehicle manufacturers and the diesel lobby to soften or slow down air pollution regulations – and manoeuvre around the regulations that do exist. These groups know what they're doing: lobbying against the health of the people to protect their own financial interests. And some car companies have even committed systematic fraud to hide the truth about their diesel emissions. The ruling on Ella's death should lead to a reckoning for these firms.

'It's now down to those in power to make up for lost time and lost lives. The government's promise to ban new petrol and diesel vehicles by 2030 is a good start, but only half of what's needed. It needs to make our streets safe to walk and cycle on, and reduce the cost of public transport, which continues to be unaffordable for the people who need it. The route out of the air pollution crisis won't be found with an investment of £27 billion into road building while the cost of travelling by train and bus continues to rise.

'I want to see the government enshrine the right to breathe clean air into legislation. This would mean that any future decisions made on transport, industry and other key areas take into account the damage that could be done to our lungs. We also desperately need legal limits and targets that are fully aligned with the World Health

Organization's recommendations. Crucially, we need to give councils the serious financial backing they need to put these plans into urgent action. Local campaign groups like the one I am involved in have sprung up around the country to push politicians to take action.'

Anjali isn't merely asking for clean air, which would be a valiant fight in itself; she is providing a road map and offering her assistance. She is making governmental inaction preposterous. Her friend was taken by a silent killer, and so she is speaking louder than ever, and many are joining her every day.

Now here is my confession. I am a recovering misanthrope. It persists in my memory so enduringly because of how much my beliefs have changed. Early on in my environmentalism, still gripped by how complex and frighteningly large the challenge was, I remember often saying to my parents how much easier it would be if all humans just disappeared. Maybe reading too many dystopian novels was to blame, but I would walk through the city and imagine great roped branches and thick leaves growing out of the buildings. Emerald lakes, grasslands and dense forest would push up from the pavements and swallow the splurge of humanity's grey concrete world. Wild megafauna would amble up Oxford Street in this vision I had of a future without humans.

The cost of this daydreaming was always the moment of return, when I had to realign my imagination with the world in front of me, stained by concrete, tarmac and steel. For me, environmentalism was about other animals, not people. I have since learnt that this is not characteristic of an environmentalist, rather a cynical mindset that causes us to romanticize roaming across landscapes devoid of people.

I've also learnt a key ecological fact: no species is superfluous. When someone says the word 'environmentalist', the first image that most people have in their minds is a sort of tree-hugging, green-loving, mung-bean-eating being. Whilst there's nothing wrong with eating mung beans and hugging trees (this has actually been proven to increase levels of oxytocin similar to those experienced when you hug a human), this narrow view of what it means to care about our planet will push people away from the movement and do more damage than good. At its core, environmentalism is about protecting life.

The planet is fine. It's been here four and a half billion years. It's been through periods far warmer and far colder than right now. It has endured five mass extinctions and would simply keep on spinning whether *Homo sapiens* walked on its surface or not. However, it's the thin crust on the top that we're trying to save. We must protect the infinitely fragile and yet startlingly resilient web of life on earth, and humans, whether we like it or not, are merely another strand of that web. Whilst we've been distracted by vanity and pride at our inventions, writing, paintings, movies and achievements, we're really just another species trying to survive, procreate and pass on our genes. Therefore, to be an environmentalist is, by its very nature, to care deeply about all animals, including humans.

Perhaps you knew this before, but now that you definitely do, it's time to introduce you to intersectional environmentalism. The phrase was coined by Leah Thomas, a young environmentalist. She defines it, briefly, as an inclusive version of environmentalism. One that advocates for the protection of both people and the planet, and acknowledges the ways that injustices being experienced by both marginalized communities and the earth are interconnected.

Leah grew up in Missouri. She was the type of young girl trembling with a desire to clamber up every tree, with wilderness seeping out of every pore, mud under every nail, a toad in one hand and a spade in the other. I was also that girl, as many young people are, but as you grow up and watch your fascination with the wilderness being leached out of you by society, you feel ripped between two paths: the stereotypical one of make-up, shopping and more 'feminine' things; and the path of trying to cling on to that wildness. Later, I realized that you don't have to fall into either camp; you can be a beautiful, messy concoction of both, and many other things too.

For Leah, being a young nature-loving, toad-obsessed girl was ostracizing enough, but it wasn't her boldness and passion for nature that made her feel alone; it was her blackness. Words were often thrown at her like knives, words that would cut deeply into any young girl, any human. Leah says, 'I was one of five black girls in my class and that experience has had one of the biggest impacts on my life.' That was where she learnt to stand up for what she believes in.

A few years later, the summer after her first year of college, Leah returned to St Louis, Missouri, where she decided to study the environmental science and policy programme at her university. As the holiday was coming to a close, she began packing up, excited for the future stretching out before her. But as often happens in life, things changed, fast, and the path she had been constructing for herself was derailed like a train changing tracks.

As she was leaving for university, folding, arranging, saying her farewells, an unarmed black teenager was shot near her house. The death of Michael Brown caused civil unrest, which flooded the news. Leah attempted to process the trauma of what she was seeing as protests erupted around her, causing literal and metaphorical fires to burn corrupt structures to the ground. She also began to examine her identity as a black woman in new ways.

She flew back to California and started her environmental science classes. She penned notes and typed lectures. She sometimes fled to the mountains or the beach to try to find meaning in the events she had witnessed. 'As my textbooks encouraged me to protect public lands so they could be preserved and enjoyed, I couldn't help but wonder, "For whom?" I started going on hikes to process what was going on and I realized nature could be a tool for healing.'

The harsh words from children at school and the even harsher actions on the streets of her home town didn't make Leah cynical or scathing in her activism. On our call, she emphasizes that no one is perfect in what they post, what they say and how they act. She rose up from the harsh words, she rose up rooted, like a tree, fully aware that to create meaningful change, you don't knock others down, you lift them up. 'I'm not saying everyone suddenly needs to transform into a racial justice radical advocate, I'm just saying you should approach those issues you care about with intersectionality. It strengthens your message. Many environmentalists are familiar with the "what" but not the "who". They know the science behind it, the causes driving it, but not necessarily the stories and people being impacted the most.'

Leah was catapulted into the spotlight on 28 May 2020, when she shared a post to Instagram defining intersectional environmentalism. The powerful pledges in her Instagram post are in the manifesto section at the end of this chapter. She gave people permission to care. She provided those who felt bound to a specific fight, and those who saw human issues as separate from environmental issues, with a gateway into the often gated environmental world. The world was reeling from George Floyd's murder, and environmentalists left, right and centre were stepping back from their platforms so as not to detract from the gravity of what was happening. In her post, Leah assured people that issues do not swell like tides, where one is important one month and another the next month. Rather, they are like drops of water all combining to make the tide itself, and so environmentalists began to emerge from the crevices and recesses of the internet and find ways to move forward in a new, more holistic manner.

Leah's model of intersectional environmentalism was based on a framework of intersectionality designed in

1991 by scholar and lawyer Kimberlé Crenshaw, who recognized that our social and political identities overlap and create different kinds of privilege and discrimination. Therefore, different types of destruction, discrimination and prejudice overlap and intersect with each other.

Crenshaw was fighting a legal battle against a sexual harassment case, and as she left the Capitol Building in Washington DC one October evening, she felt drained and dejected by the case. She saw a group of African American civil rights activists outside the building and began walking towards them, thinking, 'Oh, thank God, a place we can go and embrace each other, because this is a struggle.' Upon reaching them, however, she was struck by their T-shirts, which were proclaiming support for the very man she had just been fighting against in court. 'It was like a horror film,' she says. 'You think you're safe, but it turns out that the people you're running to are actually infected with whatever you're running from.' So intersectionality is not about fighting the separate strands of inequality, but rather addressing the very roots of power imbalances, and finding the tool by which those imbalances can be eliminated altogether.

Therefore, feminism and environmentalism are just as intimately connected as racism and environmentalism are. These strands of activism must overlap if we are to protect both people and the planet. So many of the young people I've spoken to whilst writing this book have talked about their grandmothers and mothers going to collect water, farming the land, being affected by natural disasters. A teenage girl – Greta Thunberg – began the youth climate strikes. Women are fourteen times more likely to die during natural disasters, and have been estimated to represent almost 80 per cent of the membership in environmental organizations. They are statistically more likely to take actions such as turning vegan, recycling, and reducing energy consumption. Studies have shown associations between green and sustainable products and womanhood. Up until very recently, every aspect of being a woman converged at the beating heart of femininity, which was forming and nurturing a new life. In other words, if environmentalism is caring for life, women have been perceived as environmentalists for as long as we've walked the earth. We've inherited this idea that femininity means being selfless and caring. But empathy and care shouldn't be what it means to be a woman; they should just be what it means to be a human. It's these ideas that contribute towards the gender gap, and there, in that deep abyss, many intersections lie.

It's not just racism and feminism that are incorporated within intersectionality; it includes every issue stemming from systems of oppression and dominance that destroy and damage humans and the world around them. I may have made you feel even more overwhelmed than you did before; all these issues and complex, thorny challenges make us want to retreat to a softer, more compassionate world. We want joy, laughter and love, and so knowing about all of this can turn people from anger to anxiety and apathy. As these worries bite into your conscience, though, be comforted by that ache and confusion – it means that you're human, you are the very fabric of what keeps the dream alive for a better world. Allow that worry to channel the insurrectionary current towards action.

A few weeks after I spoke to Leah, in the shifting shadows and lifting mist of a December morning, I was trying to thread all these thoughts together in my mind. There was a man sitting by his tent on the banks of the Thames. It was the same tent that had been there for months; ripped by time, aged by weather, with its flaps slapping violently. For some reason, the man had slipped under the restless gaze of the police, who usually hustle the homeless on. As I passed, his calloused hands flicked a cigarette butt into the river, then he swung his arm back and pitched his plastic coffee cup in too. Such a small, inconsequential action. The arc of an arm, the garish yellow of the butt bobbing down the river. Not unusual, but still something I'd usually condemn. But when you're haunted by the thoughts of long days of hunger and the cold winter stretching ahead, you're just thinking about surviving each day. Your mind isn't worried about melting ice sheets, future generations, and the hollow promises of politicians, because to be able to care, to act and be vocal on these issues is a huge privilege, and not one that we should automatically assume everyone around us has. This doesn't mean that you shouldn't speak up if you can, or amplify the voices of those discussing topics like racism, feminism and environmentalism from perspectives and backgrounds that aren't your own.

Make sure you grasp every opportunity to speak up. Your voice might tremble, and perhaps you'll stumble, stutter and sometimes say the wrong thing, but it's better to be an idealist with a wide-open heart trying to do the right thing than carry the burden of knowing that you could have made this world just a little bit better but chose not to.

Before I suggest how you can be one of the people who chooses to make this world just a little bit better, I want to tell you the story of someone who has done just that. Mya-Rose Craig (she also goes by the name Birdgirl) is a friend of mine from the campaigning world. As a toddler, she was strapped onto her parents' backs and taken along the shoreline of the Chew Valley Lake, near Bristol, to spot herons beating across the sky like gothic angels. At eleven, she started her blog BirdgirlUK, to document the birds she saw on her travels. At fourteen, when her sister Ayesha had a baby and couldn't come birding, Mya decided she wanted to meet other young naturalists like her. She created her organization, Black2Nature, to see more people from ethnic minority communities exploring nature and enjoying the countryside. At first, no one signed up. Now Mya has hosted many weekend-long events in the Chew Valley and the Somerset Levels, introducing hundreds of BAME children and teenagers to the natural world.

At eighteen, Mya is the youngest person to have seen half the world's species of birds (that's more than 5,000!). In 2020, she also became the youngest person in Britain to receive an honorary doctorate, which was conferred on her by the University of Bristol in acknowledgement of her efforts.

The natural world gave her a sense of belonging, so now she seeks to pass that on to others. 'Environmentalism is a very homogenous sector,' she says. 'There are a lot of stereotypes around a certain type of person that will work in the nature industry, which is being very white, being relatively middle class, and privileged.'

When you think of a birdwatcher, the image that springs to mind is perhaps a greying man in tweed, with binoculars in one hand and a pork pie in the other (maybe the pork pie is just my experience tainting my observations). With their sharp eyes and identification skills, birders are finally noticing that 93 per cent of their number are white, and that the 'racialization of space and the spatialization of race' (as author Carolyn Finney put it) means that birding and nature watching have become exclusive hobbies with invisible membership cards. If you don't see someone like yourself performing an activity, you're much less likely to perform that activity yourself.

Monocultures in farming aren't as productive and sustainable as polycultures. Monolithic cultures don't survive. Multifaceted people are the best kinds of people. Peer into a forest canopy, and you will usually see that a single flock of birds is actually made up of many different species; up to fifty species may travel as a unit to decrease competition and reduce predation. Therefore, in the environmental movement, having a range of voices is the only way to break out of this sphere we have created. As always, the answer lies in diversity.

*Manifesto for change**

- Acknowledge environmental racism and injustice. Start by educating yourself about who is most impacted by environmental injustices and encourage conversations around the disproportional impact many people face when it comes to exposure to environmental issues like poor air/water quality. Also, speak out about the barriers that many marginalized individuals face when simply trying to access nature, or the environmental space. Champion organizations like Black2Nature or Black Girls Hike UK CIC, who are working against this to create change. There are many parallels between movements that can guide us in making better decisions. For example, the civil rights and environmental movements can work side by side instead of always being separated. Social justice includes environmental justice, and environmentalism should care about people just as much as it does the planet.
- Amplify unheard and under-represented voices in the environmental space. Leah Thomas explains: 'Instead of saviourism ("How can I save these people?"), an intersectional advocate

* Based on Leah Thomas's suggestions.

asks, "How can I use my privilege to amplify the work already being done?" Chances are, there are local activists and advocates who are already fighting against environmental injustices in their communities, or non-profits doing the work. You can support these efforts by following activists and organizations on social media and sharing their content; signing petitions and supporting monetarily; and voting for local, state and federal officials supporting environmental justice.'

- Make space at the table. Describing groups as 'climate vulnerable' disempowers them, as it suggests they need to be rescued. Inclusivity requires the deconstruction of certain language. Countries and communities can have exactly the same physical vulnerability to natural disasters, but the resources and infrastructure they have access to determine how much they'll be affected and their ability to recover. Rather than making assumptions at the exclusive decision-making table about what people need, we should open up the conversation and make space so that local communities can join and offer their valuable knowledge (at panels and conferences such as COP), alongside big companies and government leaders. Especially as these individuals are usually the ones directly affected by the issues being discussed. Don't let decisions be made without them.

- Join campaigns for the government to enshrine the right to breathe clean air into legislation. It's crucial that decisions made on transport, industry and other key areas take into account the welfare of people too.
- Join and support campaigns for legal limits and targets that are fully aligned with the World Health Organization's recommendations. For example, Choked Up.

These last couple of years have proved that we *can* do things differently when faced with a crisis. We should harness that energy in 2022 and beyond. A national network of stricter clean air zones and school streets to protect children from traffic pollution would be a good start. As would a national public campaign to make sure that parents are never again deprived of the information they need to protect their children: the coroner highlighted a failure to provide Ella Kissi-Debrah's mother with information about the way air pollution could exacerbate asthma. We must work together to create a world where this cannot happen.

Leah Thomas's intersectional environmental pledge

I will stand in solidarity with Black, Indigenous + POC communities and the Planet

I will not ignore the intersections of environmentalism and social justice

I will use my privilege to advocate for Black + Brown lives in spaces where their messages are often silenced

I will proactively do the work to learn about the environmental and social injustices Black, Indigenous + POC communities face without minimizing their voices

I will respect the boundaries of BIPOC friends and activists and not demand they perform emotional labour or do the work for me

I will share my learnings with other environmentalists and my community

I will amplify the messages of Black, Indigenous + POC activists and environmental leaders

I will not remain silent during pivotal political and cultural moments that impact BIPOC communities

Conclusion

Now It's Your Turn . . .

Picture this: Greta Thunberg on a train, the softened blur of a passing forest framed by the window. The table in front of her is spread with reusable water bottles, paper cups, vegan snacks and, yes, a few pieces of single-use plastic. This scene describes a photo Greta posted on Instagram in January 2019. You would think, given her monumental impact on our collective conscience, and the stratospheric nature of her emergence onto the world stage, that people would forgive this small, almost inconsequential part of the picture. But no, she was derided and castigated; the photo spread like wildfire and was turned into ammunition by critics, who used it to highlight 'hypocrisy' and suggest that somehow, because one teenager used two pieces of plastic, the entire environmental movement was a scam.

This is a big problem. Not the plastic in this case, but the call-out culture, the notion that if you wish to speak out about something, your life must be purged of hypocrisy. I'll tell you now before we go any further: I'm a hypocrite, and it's impossible not to be. We live in a society designed to make us buy more, spend more, eat more – and think and care less about the impact we're having on the environment. Salvation from this climate crisis will not come in the shape of a reusable cup, or a bag for life. Our bamboo spoons won't dig us out of the pit of environmental damage. We are all firmly embedded in a broken system, and so the system must change.

However, it goes without saying that any system is made up of components. That's us. We are the building blocks, the consumers, and although one individual action won't save the world, the accumulated actions of a unified community can impact political, social and economic structures. If we work together, things will happen. We can be the catalysts of structural change.

This chapter, on how to act, is not teaching you to be an activist in the conventional sense of the word. When you picture an activist, megaphones, placards and podiums come to mind, all tools to elevate and amplify, to make you as loud and unmissable as possible. But to be effective and speak so that people *listen*, you don't necessarily have to make more noise. Some of the loudest actions have been committed by the quietest people; Mahatma Gandhi, Rosa Parks and Eleanor Roosevelt were all notorious introverts. So dissolve whatever images you have in your mind of what you must be or do to make change, and together we'll look at some of the most useful lessons I've learnt so far, through other campaigners and people trying to make change in this world, however big or small.

Find your why

Studies have shown that people who feel as though they lack purpose in life are more likely to have worse mental health. This feeling also negatively affects their mood, relationships, and ability to think and reason effectively. When you find your why, you'll know. When it happens, you'll feel like you've brushed up against a live wire and suddenly been charged with a knowledge of what you must do. You don't always know how to do it, but that is part of the process. Activism is imbued with self-righteousness, and you might feel like everyone else has it all figured out, but sometimes taking that first action itself is where you discover your niche. Educate yourself and stand up for what you see as being the most compassionate thing to do, always, and somewhere along the way you'll discover your why.

For some help on this process, watch the TED talk 'How great leaders inspire action', by Simon Sinek.

Restructure your environment to sustain your own green habits

The primitive side of us wants to save as much energy as possible, and we like it easy. Behaviours that require less friction are more likely to become habits, whilst willpower alone rarely creates habits. This is why many New Year's resolutions fail. Most big companies have mastered the art of tapping into this part of the human psyche. Binge-watching on Netflix or Hulu is facilitated by the way an episode starts immediately, meaning we don't have to go and find the next one. The app for Uber is actually designed to minimize friction. Your phone's GPS knows where you are, and you don't have to waste time handing over cash. Just a few clicks and you're done. In a study where one group had to commute 3.5 miles to the gym, they were likely to go five times a month, whereas the group who had to commute an extra mile were only likely to go once a month. In other words, me presenting you with all these possible behaviour changes in this book doesn't mean much unless you make sure they become habit and you reduce the friction for yourself to carry them out regularly. Do your research now on sustainable brands so you don't have to check every label when you go to the supermarket. Leave your bike tied up right outside your house, not trapped away in a garage, and you'll be more likely to cycle to school or work in the morning. Subscribe to newsletters that will deliver environmental headlines to your inbox and keep you updated without you consciously having to go on a hunt for news.

Be firm, not neutral, but don't create impenetrable boundaries

People who step up and take action, whether child or adult, often experience vitriol and personal attacks. I see social and environmental movements a bit like boats. When a boat is travelling faster, it faces more resistance from the water. When movements are gaining momentum and having an effect, they too face more resistance, as they're reaching people in corners and crevices that previously haven't been accessed, creating discomfort as others begin to recognize the flaws in how they're living. Initially,

I thought neutrality and trying to pacify opponents by agreeing with them to some extent was the key. It's not, but neither is completely dismissing them. You are probably not always right, definitely not, in fact, and therefore you should always remain open to other ideas and be ready to change your mind, even if it challenges your beliefs and self-righteousness. However, when you're standing up for what you think is the right thing to do, remain compassionate and open even to those who disagree – and remember, you don't have to validate their arguments. If you consider them, and decide you don't agree, you can let them know that. Deep down in that other person, however convinced and impenetrable they may seem, there is a conscience that is listening, and your words might send little cracks snaking through that facade.

Remember, neutrality and silence help the oppressor, never the victim. You don't have to please everyone. Always be willing to accept that you too may need to change the way you think about something.

Your voice is always needed, and you are never an imposter

This message is to myself as much as to you. I know well the feeling of thinking that you are way out of your depth, yet already so entrenched in a situation that you cannot back out. I have asked myself why I am writing a book about the climate crisis when I'm still at school. However, I remind myself that I'm not pretending to know things I don't know. I can touch on the science of it, and have done, but I know that in-depth science should be left to the specialists. I've found a place in the environmental movement where I can share the stories from those young people at the heart of the climate crisis, so that we're heard together, across our differences. You might feel like an imposter because you're not like everyone else in the movement. However, it is this quality of uniqueness and diversity that makes your voice important. Now, as imposter syndrome requires far more attention than I can give it in this short paragraph, I suggest you research the Dunning–Kruger effect. All may not be as it seems . . .

You have to make sacrifices, but don't sacrifice yourself

Jamie Margolin, the eighteen-year-old co-founder of one of the most effective international youth climate justice organizations, Zero Hour, is one of the most prominent young people in the modern climate movement. In her recent book, *Youth to Power*, she talks about her routine – getting up at 5 a.m. and then being wrenched between conferences, calls, school and protests until 10.30 p.m. – which eventually sucked her of all joy so profoundly that she became solely defined by her activism. She was an activist, and almost that alone. This has happened to many young people I've spoken to, who have succumbed to exhaustion, or burnout (a term first coined in 1974 by the German-American psychologist Herbert J. Freudenberger).

Burnout is dangerous for you as an individual, but also for the movement and the change you want to create. Paul Gorski, who has interviewed hundreds of activists, found that roughly half of those who reported experiencing burnout didn't take time off – they simply left their movements for good. So, rather than seeing activism as a moment in time, or a trend, focus on the movement. Movements have peaks and troughs, so allow yourself time for rest and recuperation if you're in this for the long run.

I've seen and experienced the culture of 'martyrdom' almost intrinsic to social movements. At the age of twelve, I began doing media interviews with no idea *how*, yet fully aware of my *why*. After a few years, I knew *how* to conduct myself in those situations, but somewhere along the way my *why* had become obscured by the activist persona I'd shimmy into whenever I was on the news or on a stage. I would repeat the same statistics, the same desperate pleas and statements, and rather than digging into my passion to power my words, I was digging into past interviews and regurgitating them. Don't ever become so encompassed by the activist persona that you detach yourself from the old you who began the action. Instead, keep finding ways to nurture your passion for change. For me now, this is being out in nature, or finding creative ways to express myself.

Find joy in making change, or you won't sustain it

The end result shouldn't always be the yardstick for success. The process, the connections made and the groundwork laid are equally as important, for you are more likely to continue if you find joy in the process. This joy will make sure you do not retreat into cynicism or settle into disaffection.

Very few people take part in activism because they enjoy it, but rather because they feel some inexplicable desire to have an impact, to make things just a little bit better in any way they can. However, in order to sustain (and enjoy!) what you're doing, incorporate yourself and your passions into the process. Personalize it by playing to your strengths, add nuance and context. It's only when everyone has a role to play that the movement will become as widespread as possible. Lesein Mutunkei, for example, who I interviewed for Chapter 7, found a way to merge his love of football with his passion to protect the environment. Dara McAnulty, who relishes the soft glow of a gathering twilight and the comfort of putting pen to paper, shares his love of nature through the written, rather than the spoken, word.

If you enjoy cooking, help others transition to a plant-based diet. Maybe start a blog, or write a cookbook. If you're a lawyer, or want to become one, consider specializing in environmental law. If you're a parent, teach your children how to be aware of the impact they're having. If you're a teacher, show your students. If you're at school, mobilize those around you. Don't try and fit into the mould of what an activist should be. Instead, find your patch and cultivate it like a garden, rather than trying to water someone else's.

Find the community

To achieve large-scale change, there's no such thing as 'just individuals'. The very fabric of change requires stitches sewn by many, many people at a systemic level.

Individualism is so pervasive in Western societies that this discourse has trickled through into activism. The idea of the problem resting upon the consumer's shoulders was popularized in 2004, when BP produced a 'carbon footprint calculator' so individuals could assess the impact they were having, making them feel guilty. According to Benjamin Franta, who researches law and history of science at Stanford Law School, it was 'one of the most successful, deceptive PR campaigns maybe ever'. In this perfect storm of industry, culture and psychology, big polluting industrialists want individuals to feel guilty about pollution, in the same way tobacco companies blamed smokers for overusing addictive carcinogenic products.

However, although we're being led to believe that the most we can do is recycle our plastic and hop on a bus rather than take the car, no one single person is the answer to this crisis. Some may point to Greta Thunberg to show how one individual can have a huge impact, but the impact itself was the mass mobilization of young people worldwide, which, clearly, relied upon millions. A cult of personality puts too much pressure on one person, imprisoning them in a role so they cannot take a step back, whilst also omitting other voices that are crucial to the movement's success.

Don't be afraid to reach out to others; ask for advice and mentorship and work collaboratively. Build upon the foundations others have put in place, rather than trying to start from the beginning, or each generation will go back to square one.

Crucially, know that you, as an individual, have an incredible amount of power to effect change in your own life. However, it's when we reach out, work with and uplift others that the actions coalesce and accumulate, allowing us to get to the root of the problem.

Be creative. Allow yourself to imagine!

This predicament we find ourselves in now is an opportunity to invent another kind of civilization. One more cognizant of limits and less exploitative, but also one more compassionate, green, inclusive and safe. Just as volcanic ash creates the most fertile soil, we too now have the opportunity to exit this period of smothering the earth and allow for the growth of something else. Therefore, when you decide to take action, spend time envisioning what you want the world to look like. What is the dream of a better society/cleaner environment/more compassionate world you want to move towards? The reason this is important is akin to approaching someone and telling them they are going in the wrong direction without telling them the correct path. We can't denounce the dominant narrative of overconsumption and destruction of all that is beautiful unless we create a powerful and compelling alternative. I'm not religious, but I do think that as a species we have to believe in something bigger and more potent than ourselves. Something that sculpts our morals and drives our actions. Unity is crucial. All change that has ever been created, and that will ever be created, has begun rooted in the individual or collective imagination.

Just as humans are moulded by the landscape around them, the external world, the environment, is moulded by our internal landscape. There is a landscape outside the self: the stern curve of a mountain slope, the viscous air of gathering dusk, the bubbling of the dawn chorus, the bent trunk of a wind-blown tree. Then there is the landscape within: the way we see the world and the way we want to see the world going forward. It is only when we change this internal landscape and see nature as integral to our well-being and survival that we can protect the external. How can we look after the environment if we're not first connected more deeply to it as humans? I used to wonder how much the land shapes the imaginations of the people who dwell in it. Maybe that's why different cultures see the world in so many different ways. Now, instead, we must focus on 'How much does the imagination of the people shape the land they dwell in?'

Activist highlight

Kehkashan Basu has inspired thousands of young people around the globe. She is already an author, Climate Reality mentor, TEDx speaker, and the founder president of social innovation enterprise Green Hope Foundation, which provides young people in 16 countries with a networking platform to engage in the sustainable development process. She has spoken at over 155 United Nations and other global forums across 25 countries and has won more prizes than my word count would allow me to list.

Boxing Day 2020. It has been a disorientating year in a tumultuous world, and yet Kehkashan bubbles onto the screen, with ocean-blue slashes of eyeshadow streaked across her eyelids, smiling and determined. When I ask her how she spent her Christmas, she says, as casually as if she was telling me what she had for breakfast, that she spent the day installing solar panels in Liberia as a Christmas present to the local communities, so that the children would be able to study using proper light rather than kerosene lamps. This has been proven to improve their chances of further education, whilst also ensuring they do not descend into drugs and crime. It's symbolic, really, of how an action as simple and sustainable as installing solar panels doesn't just reduce energy use and therefore benefit the environment, but also impacts on many other areas of people's lives too. Again, it's that idea of intersectionality coming into play.

As a species, we have a bad habit of classifying things and people into binary systems. Orwellian or utopian, male or female, pessimist or optimist, systemic or individual, active or passive. Kehkashan is one of those people who politely but firmly karate chops apart any boxes you try to fit her into. She's not an environmental activist, she tells me first.

'Do you believe more in individual or systemic change?' I ask cautiously, knowing that many people hate this question, yet for some reason always feeling compelled to ask it.

'They go hand in hand. The system is run by people, and if the people start by taking individual actions, they learn how to have the power to bring about systemic change. I firmly believe in both; only then can we achieve a sustainable world.'

It's this idea of sustainability that drives Kehkashan, and is perhaps the core lesson I've learnt speaking to so many young people on such a wide range of issues.

Kehkashan was born in 2000, on World Environment Day. When she was seven, she saw, in a magazine, an image of a dead bird, head lolling backwards, stomach brimming with plastic, and was hit suddenly by the garish light of reality. However, although she was tender enough to be distraught at what she saw, she was tough enough that some burning desire for change began to unfurl within her. Around the same time, she attended a lecture by the great explorer Robert Swan, who had just got back from an expedition in Antarctica, and who said that 'The greatest danger to our planet is the belief that someone else will save it . . . The last great exploration on earth is to survive on earth.' And so this little girl became one of the great explorers of the twenty-first century, on a quest to save humanity from itself.

At the age of twelve, she founded the Green Hope Foundation, and was then elected for a two-year term as the UN's Global Coordinator for Children & Youth and a member of its Major Groups Facilitating Committee. This isn't about awards or accolades, though. There is no 'hero' cloth that Kehkashan is cut from. The difference between her and someone who hasn't stepped up in the way she has is the mindset she has built. It is so easy for us to personify good and evil – almost inevitable, actually. Creating a dichotomy of 'us' as the good ones and 'them' as the bad offers us comfort, as we feel that there is someone to blame, cancel and defeat. Kehkashan doesn't do this.

I ask why it's important for young people (commonly seen as the victims) to be involved in fighting to protect the environment. 'Don't just see young people through the lens of their age; see them as powerful because of what they've done. You see a problem, you solve it. It doesn't make you one of the "good" ones, you are doing what needs to be done. Holding others accountable and playing the blame game means nothing unless you also hold yourself accountable.'

Speaking to Kehkashan was the last interview I did for this book. I have learnt so much from these young people and the ways they're taking action to create a better world. We've journeyed to desecrated lands that are being nurtured back to life, ventured into the wilds of the jungle, scaled mountains and hopped continents. We've been here to *listen*, always listening, to the stories from the heart of the climate crisis that need to be heard. Yet all the problems we've witnessed and all the lives entangled in the mess of destruction we've created cannot be corrected by just picking away at the individual knots. We must first stop whatever caused the knots in the first place, which, as all the young people have told me, is us. As Kehkashan says, don't confine yourself to the label of 'youth', 'adult', 'perpetrator' or 'activist'. Great hope lies in the fact that whilst we are the problem, we are also the solution.

I hope you read the words and stories in this book and feel inspired to grab them with both hands and use them to help protect people and the planet. Writing, I believe, is a means of drawing from the well of your own beliefs and sharing the powerful stories of others, whatever they may be. At the heart of this book is the simple, abiding belief that another way of living more gently and more wisely on this earth is possible, and when we all commit to searching for that and making it a reality, I imagine this stifling carelessness falling away from us.

We've been straining to escape the animal world, closing ourselves in and locking nature out. Now the repercussions of our actions are pulling us back into it. Either we live apart from nature and continue on this trajectory, or we learn to live within it, and let it live within us too.

Acknowledgements

To my mum, for never quelling my idealism and passion when I wanted to bring snails into the house at four, create an orang-utan sanctuary at ten, shut down the fossil fuel industry at twelve, and then – perhaps craziest of all – write a book. To my dad, for always encouraging me to question deeper, explore further and learn more. To Emily Robertson and Susannah Bennett at Penguin for believing in me and providing me with an opportunity to fulfil a dream and spread a message. To Cyril Dion, Baptiste Morizot and the crew of *ANIMAL* for embarking on the journey where I learnt the importance of narrative and listening to the stories of others, and the power that can wield. To all the young activists who uplift each other, who turn despair into action rather than apathy, and who care deeply, so deeply, about life on this planet. To all those I've met who care passionately, fervently whilst remaining empathetic and self-effacing enough to bring others with them. You inspire me every day.

Bibliography

Our disconnect from nature

http://www.forestpeoples.org/sites/fpp/files/publication/2010/08/cerdcostaricacerdurgentactionjul10eng.pdf

https://www.theguardian.com/environment/2020/feb/25/costa-rican-indigenous-land-activist-killed-by-armed-mob

Women and climate change

https://www.worldvision.org/clean-water-news-stories/compare-walk-for-water-cheru-kamama

https://www.theguardian.com/global-development/2015/dec/10/women-injustice-climate-change-thoughts-from-the-paris-talks

https://ideas.ted.com/want-to-fight-climate-change-educate-a-girl/

https://books.google.co.uk/books?id=ooD7CwAAQBAJ&pg=PT145&lpg=PT145&dq=african+woman+travelling+further+to+collect+water&source=bl&ots=WFMbdlYLP_&sig=ACfU3U1wc4epc_WtjeI5_eVIiuRWqwSw6w&hl=en&sa=X&ved=2AHUKEwi-07nooprqAHXSVRUIHe50C-8Q6AEwC3OECAoQAQ#v=onepage&q=african per cent20woman per cent20travelling per cent20further per cent20to per cent20collect per cent20water&f=false

https://www.internationalwaterlaw.org/bibliography/articles/Ethics/Women_and_Water.pdf
https://en.wikipedia.org/wiki/Ogbomosho

Stages of life development

https://www.simplypsychology.org/Erik-Erikson.html

Autumn Peltier

https://thetyee.ca/Opinion/2019/09/30/Autumn-Peltier-Asks-Why-So-Many-Communities-Have-No-Water/
https://www.humansandnature.org/videos#sb=https://www.youtube.com/watch?v=zwStzuCwRtM

Plastic

https://www.nationalgeographic.co.uk/2018/05/we-made-plastic-we-depend-it-now-were-drowning-it

Greenwashing

https://www.countryliving.com/uk/news/a28659267/what-is-greenwashing/

Consumerism

https://www.lifesquared.org.uk/system/files/Consumerism per cent20download_1.pdf

'Is the Growth of Resale Really Linked to Sustainability?', *Vogue*, April 2018

Mumbai

https://nextcity.org/informalcity/entry/we-need-more-slums

Air pollution

https://www.researchgate.net/publication/282867136_Psychological_responses_to_the_proximity_of_climate_change
https://www.independent.co.uk/environment/european-smog-27-times-more-toxic-chinese-air-pollution-china-quality-a7572051.html

Travelling

https://www.bustle.com/p/why-do-we-like-to-travel-so-much-heres-why-wanderlust-is-such-a-strong-feeling-according-to-science-9399514
https://www.amexessentials.com/brief-history-staycation/
https://theconversation.com/is-this-the-end-of-the-road-for-business-travel-36226

Agriculture

https://www.vox.com/2014/8/21/6053187/cropland-map-food-fuel-animal-feed

https://www.theguardian.com/environment/2018/dec/21/lifestyle-change-eat-less-meat-climate-change
https://www.nationalgeographic.com/environment/future-of-food/photos-farms-agriculture-national-farmers-day/

Artemisa

https://unfoundation.org/blog/post/how-xakriaba-brazil-protecting-biodiversity/
https://www.democracynow.org/2019/9/23/brazil_indigenous_climate_activist_artemisa_xakriaba
https://www.survivalinternational.org/articles/3540-Bolsonaro

Submergence

https://www.zmescience.com/science/sea-level-rise-solomon-islands-sink/

Women and water

Water-saving tips, Ofwat
'Diet change – a solution to reduce water use?', IOPscience, July 2014

Intersectional environmentalism

'Top Oil Industry Group Disputes African-American Health Study, Cites Genetics', Inside Climate News
http://tupress.temple.edu/book/0701

Leah Thomas, Intersectional Environmental Pledge, Staging Change

Now it's your turn

'People Age Better if They Have a Purpose In Life', *Time*, August 2017

Index